W9-DGF-171

MAD SHEEP

MAD SHEEP

The True Story Behind the USDA's War on a Family Farm

LINDA FAILLACE

Foreword by Ronnie Cummins,
National Director, Organic Consumers Association

CHELSEA GREEN PUBLISHING COMPANY
WHITE RIVER JUNCTION, VERMONT

Editor: Helen Whybrow
Managing Editor: Marcy Brant
Copy Editor: Janet Jesso
Proofreader: Eric Raetz
Indexer: Peggy Holloway
Designer: Peter Holm, Sterling Hill Productions
Design Assistant: Daria Hoak, Sterling Hill Productions

Printed in the United States
First printing, August 2006
10 9 8 7 6 5 4 3 2 1

Our Commitment to Green Publishing
Chelsea Green sees publishing as a tool for cultural change and ecological stewardship.
We strive to align our book manufacturing practices with our editorial mission and to
reduce the impact of our business enterprise on the environment. We print our books and
catalogs on chlorine-free recycled paper, using soy-based inks, whenever possible.
Chelsea Green is a member of the Green Press Initiative (www.greenpressinitiative.org),
a nonprofit coalition of publishers, manufacturers, and authors working to protect the
world's endangered forests and conserve natural resources.
Mad Sheep was printed on Ecobook 100 Natural, a 100 percent post-consumer-waste
recycled, old-growth-forest-free paper supplied by RR Donnelley.

Library of Congress Cataloging-in-Publication Data

Faillace, Linda, 1965-
  Mad sheep : the true story behind the USDA's war on a family farm / Linda Faillace.
    p. cm.
  Includes index.
  ISBN-13: 978-1-933392-09-7 (hardcover)
  ISBN-10: 1-933392-09-6 (hardcover)
  1. Sheep–Vermont–History. 2. Sheep–Diseases–Prevention–Government policy–United
States. 3. Faillace, Linda, 1965-   . 4. United States. Dept. of Agriculture. 5. Bovine spongi-
form encephalopathy.   I. Title.
SF375.4.V5F35 2006
636.3'089683--dc22
                                    2006014225

Chelsea Green Publishing Company
P.O. Box 428
White River Junction, VT 05001
(802) 295-6300
www.chelseagreen.com

CONTENTS

*In search of love, abundance, happiness, and healing,
this was written for
Larry, Francis, Heather, Jackie,
and our beloved flock of sheep.*

When the government fears the people,
you have liberty.
When the people fear the government,
you have tyranny.

THOMAS JEFFERSON

FOREWORD

Warning: reading this riveting tale of good and evil will make you angry. After a few chapters, you will likely suffer from a deep sense of disillusionment and an uncontrollable urge to speak out or strike back. As this twisted and wicked pastoral unfolds, outrage after outrage, it will become increasingly clear that this is not just a tragic case of administrative bungling or faulty science on the part of the USDA (United States Department of Agriculture).

What we are confronted with in *Mad Sheep* is a government conspiracy. A politically inspired ritual of fabricated charges, manipulated science, and doctored evidence. A modern witch-hunt to sacrifice the innocent in order to protect the massive profits and scandalous practices of the guilty. A diabolically orchestrated, media-scripted search and destroy operation in the Vermont countryside, designed not just to murder some innocent sheep and thereby exorcize mounting consumer fears about food safety and mad cow disease, but also to turn us all into sheep, to fan the flames of fear and ignorance, and to foster our continued dependence on an abusive Big Brother government that has promised to protect us from the contemporary terrors that lurk, well, nearly everywhere.

In their highly acclaimed 1997 book, *Mad Cow USA: Can the Nightmare Happen Here*, John Stauber and Sheldon Rampton recount how they used a Freedom of Information Act investigation to pry loose secret planning documents from the beef industry and the USDA. These liberated documents included a "crisis management" plan for how to manipulate public perceptions and concerns surrounding a likely outbreak of mad cow disease in the United States. A central part of this plan was to keep quiet about the fact that leading experts on mad cow disease, such as Dr. Clarence Gibbs and Dr. Richard Marsh, had warned government officials as early as 1989 that a form of mad cow disease was likely circulating in U.S. cattle herds, and that the extremely risky, profitable practice of feeding blood and slaughterhouse waste to animals needed to be halted immediately.

By the late 1990s, as scores of Europeans who had eaten contaminated beef from mad cows began dying from a fatal brain-wasting human disease called new variant Creutzfeldt-Jakob Disease (nvCJD), USDA officials began to implement their crisis management plan. As wary Americans began to learn about mad cow disease and animal cannibalism and cut back on their beef consumption, something had to be done to restore consumer confidence. Since stopping the feeding of animals to animals and ordering universal testing of cows for mad cow were deemed serious threats to industry profits, the USDA's public relations specialists came up with diversionary tactics: round up imported livestock across the country and harass family farmers like the Faillaces, all in the name of preventing mad cow disease in America.

Meanwhile the disinformation flacks at the USDA, aided by public relations firms and the news media, worked to sweep under the rug the alarming fact that U.S. corporate agribusiness was doing exactly the same thing that Europeans had been doing to spread mad cow disease—*feeding cattle, pigs, chickens, households pets, and deer and elk on game farms billions of pounds of blood, slaughterhouse waste, animal fat, and tainted manure every year.*

The Faillace's sheep were absolutely healthy and presented no risk whatsoever to the American public. But the Clinton and Bush administrations wanted a scapegoat. Never mind that the imported sheep on the Faillace farm had been quarantined and certified by European Union authorities as having never consumed slaughterhouse waste. Never mind that these sheep had been tested for scrapie, so-called mad sheep disease, and found healthy. Never mind that a breed of imported sheep on the Faillace farm, East Friesians, has never in recorded history suffered from a single case of scrapie. Scrapie has been endemic in U.S. sheep herds for decades, and as Stauber and Rampton point out in *Mad Cow USA*, the USDA has done little or nothing to help U.S. sheep farmers eradicate the disease.

Blame Europe and a small group of stubborn Vermont sheep farmers for endangering public health, and maybe people wouldn't notice that American agribusiness had continued importing hundreds of millions of pounds of bargain-priced slaughterhouse waste from the UK for eight years after the outbreak of mad cow disease in 1989. Maybe people wouldn't notice that the 1997 FDA "feed ban" on feeding ruminants to ruminants is full of holes (allowing cattle blood, cattle fat, poultry manure and slaughterhouse waste from pigs and chickens to continue being fed to cows). Maybe consumers would forget that Europeans have boycotted U.S. non-organic beef and poultry—routinely laced with antibiotics and hormones—since 1988.

As for those farmers and consumers who won't behave like proper sheep, who refuse to shut up and swallow the official story: harass and threaten them, seize their animals, ruin their reputations, and destroy them financially and psychologically.

Linda and Larry Faillace, and their children, Jackie, Heather, and Francis, along with the Vermont consumer and farm activists who stood by them, are not only good shepherds, they are national heroes. USDA bureaucrats like Linda Detwiler, CDC bureaucrats like Lawrence Schonberger, indentured politicians, and their puppet masters behind the scenes—the leaders of the corporate-industrial agriculture and pharmaceutical complex—are the real offenders.

After campaigning in the trenches for thirteen years to get the USDA and FDA to stop the hazardous feeding of billions of pounds of blood, slaughterhouse waste, and manure every year to farm animals, and to require mandatory testing of cattle for mad cow disease at slaughter, I had lost or repressed some of my anger and frustrations. But then I read this book, and like post-combat stress, a flood of memories rushed back.

Hate mail arriving at my Washington office in 1993 along with a series of anonymous telephone death threats to my colleagues, just after we launched a national campaign against McDonald's and filed a legal petition to stop the feeding of animals to animals. A creepy ex-military intelligence agent provocateur who infiltrated our campaign and followed me around Washington, posing as a representative from the World Council of Churches. A private investigator in Wisconsin reporting that our office telephones were tapped, probably by the beef industry. A national news producer sheepishly apologizing to me for "alterations" in the script of a nationally televised ABC News story on mad cow disease that aired in 1997—following what he described as a "call from the White House."

And more. Fruitlessly petitioning the Centers for Disease Control to make the human equivalent of mad cow disease, Creutzfeldt- Jacob Disease (CJD), an officially reportable disease. Petitioning the CDC, again in vain, to require autopsies for a significant number of the 50,000 Americans who die every year from Alzheimer's disease, to determine whether they actually had CJD (CJD is often mistakenly diagnosed as Alzheimer's, because its symptoms are similar). Watching the Bush administration USDA blame the Canadians for our first mad cow cases, and shortly thereafter threaten a Kansas meat packer, Creekstone Farms, for the "crime" of wanting to test all of their cows at slaughter for Mad Cow disease.

No wonder millions of Americans no longer trust the government or the media. No wonder millions of consumers are turning away from industrial meat and food and voting with their pocketbooks for healthy, sustainable, locally produced organic foods.

But voting with our consumer dollars is not enough. The mad sheep battle described in these pages is not an isolated case. Armed with $90 billion in taxpayer money each year, the USDA is waging war against all of us—consumers, family farmers, farm animals, and the environment. The direct and collateral damage of this war includes rampant water, air, and food pollution; an epidemic of cancer, birth defects, obesity, and hormone disruption; pollution by genetically engineered crops; an unsustainable, massive venting of climate-destabilizing greenhouse gases; pesticide and antibiotic contamination; proliferation of junk food; systematic exploitation of small farmers, farm workers, and slaughterhouse workers; and the dumping of millions of tons of subsidized crops and meat at below the cost of production on developing nations, thereby destroying the livelihoods of millions of small farmers and rural communities.

It's time to follow the example of the Faillace family. It's time to stand up and fight, not only for ourselves, but also for future generations.

Venceremos! We Shall Overcome!

RONNIE CUMMINS
Organic Consumers Association
July 2006

1

ENGLAND

It all began in England. It's funny how you can look back at your life and find distinct times that mark the beginning of events, some that took days or weeks to experience and others years. Often it is harder to see the end. But for this chapter of my life, England was definitely the beginning. We were a young, ambitious family full of dreams when Larry was offered a position at the University of Nottingham School of Agriculture in England. He had just received his PhD in animal physiology from Virginia Tech University in 1990, and we jumped at the chance to live and work overseas.

England was even more beautiful than I imagined. With a population of almost forty-eight million in an area the size of Alabama, the amount of farmland and open spaces was startling. Centuries-old rock walls surrounded beautiful rolling green hills, and white sheep dotted the pastures. It was August—sunny and glorious—and I instantly fell in love with the country.

The campus for the School of Agriculture was located in a charming rural area of the East Midlands, twelve miles south of the city of Nottingham. Larry's contract provided a house for us, but it wasn't ready when we arrived in the small village of Sutton Bonington. Since the students were still on summer break, the university temporarily moved us into a dorm—our own dorm. Our children (Francis was five years old, Heather four, and Jackie three) loved telling people we lived in a big house with twenty-two bedrooms, twelve "loos," six showers, four bathtubs, and one tiny kitchen.

"You'll just be here for a week or so," we were assured. The first week went by, then the second. During the third week the dorm began filling with students coming back for school.

"Only a few more days and your house will be ready," we were repeatedly promised.

Then one day Larry came home during his lunch break. "I have something to show you," he announced. The children and I followed him outside, across the grounds, to a row of semi-detached houses at the edge of campus. Each had a wooden fence around the perimeter, a small front yard, and a large backyard (referred to as the garden).

"Look for number fifty," he said.

We peered into the yards, searching for any numbers that would identify the houses. "Hello!" Francis called to a young couple sitting in their garden, enjoying the beautiful autumn day. They smiled and waved.

A few doors down we passed a lawn scattered with mounds of children's toys. "Not this one!" Heather yelled.

At the very end of the dirt lane stood two houses. Both were empty. The first had a nicely manicured lawn; the other looked like something from *The Addams Family*. On the latter, Jackie spotted a small sign, 50.

We pushed open the creaky back gate with its chipped paint and walked into a large garden. A few trees formed the back boundary, and overgrown plants fought with one another. It was obvious it had been weeks, if not months, since the lawn was mowed. A sidewalk, barely visible through the weeds, led us to the back door.

Larry held up a key.

I gasped, "How did you do that?" He just smiled and kissed me.

The house was deserted except for a small fridge, stove, and some dishes. None of this mattered. We preferred to sleep on the floor in a home of our own rather than stay another night in the dorm. And we did.

A few days later a large moving truck pulled up to the house. "You the Americans?" the driver asked me.

"Yes. Can I help you?"

"The university ordered some furniture for you." It was beautiful, brand-new cherry furniture (beds, wardrobes, tables, chairs) and a front-loading washer. Our dreams were coming true.

Sutton Bonington was originally two settlements that melded in medieval times and as a result had two parishes. St. Michael's was the "newer" church as it was only 600 years old, compared to St. Anne's, the 900-year-old church located at the far end of our village. We enrolled Francis in the local primary school, and Heather attended St. Michael's preschool. Too young to go to school, Jackie spent her days following me around, tugging at my skirt, asking, "What are we going to do next, Mummy?" (All the children had quickly acquired the local English dialect.) I tried to keep her busy: explaining the different plant names as we worked in the garden, putting puzzles together, playing games, and showing her how to bake cookies. One morning shortly after we bought our first computer, I opened a program entitled PC Paintbrush and encouraged her to learn how to use it and to call me when she was done. As I left the room I thought, "This should buy me half an hour to get some housecleaning done."

When Larry came home for lunch, Jackie's eyes were still fixed on the computer screen. "Look, Daddy. I can draw lines!" She had figured out how to use the pencil icon and had drawn a series of lines on the screen.

"Time for lunch," I said.

"No. I don't want lunch," Jackie stubbornly replied.

"You can't sit in front of that all day," Larry told her.

"I'm still learning," she insisted.

By that evening Jackie had a computer-generated drawing of a yellow boat with a blue sky. We were amazed. "Time to get her into school," I told Larry.

Luckily Heather's preschool was willing to accept Jackie early because the girls were very close, not only in age but as friends, and would take care of each other. This now left me with plenty of free time. I spent my days visiting with our wonderful new friends, teaching piano, walking, reading, and happily restoring the garden and discovering its hidden features: a stone patio outside the kitchen door, a rock wall around the border, and stone steps leading to an upper terrace.

As much as I enjoyed my free time, I was soon bored and began visiting Larry more often. Larry was hired to study the prolificacy of Chinese Meishan pigs and teach laparoscopic artificial insemination. For him it was the perfect combination of research, teaching, and hands-on animal work. He was an incredible teacher, had a very steady hand for surgery, and was very gentle. I loved watching him work in the laboratory and never tired of our long discussions. When Larry's lab technician became pregnant and had to quit, I jumped at the opportunity to take her place. Larry spoke with the head of the department, Professor Eric Lamming. "Prof," as we called him, was a tall, handsome man in his midsixties with a full head of white hair and sparkling blue eyes. He was quick with a joke, and I liked him immediately.

"What experience does she have?" he asked Larry.

"None, exactly. But her father is a biochemist, and she grew up in his lab. Plus, she's a quick learner."

"We'll give her a try, but if she can't prove herself in two weeks, we'll have to hire someone else."

Larry ran back to his office where I was waiting. "You're in!" he told me.

My lessons started during lunch that very same day. Larry taught me about the reproductive system, the different hormones and their functions, and what he and his team were trying to measure and why. The next day was my first day in the lab, and I loved it: the equipment, the smells, and wearing the lab coat all reminded me of wandering through my father's laboratories as a child. I listened intently and took extensive notes as Larry explained the steps for measuring progesterone using a radioimmunoassay, a technique to quantify minute amounts of a hormone or drug using a radioactively labeled hormone to bind to a specific antibody. Before I realized how much time had passed, it was time for tea. "You go ahead," I told Larry as I continued to write in my notebook. "I want to work here for a while."

"You can't. No one misses morning tea," he insisted. "It would be an insult to the others." Our department was made up of an eclectic mix of scientists from England, Ireland, Scotland, and Wales. Larry explained how teatimes were opportunities to find out what our colleagues were working on, to coordinate with them, to relax, and, most important, to socialize. Sure enough, when we walked downstairs the tearoom was already buzzing.

After tea it was back to the lab. During lunchtime Larry and I would walk the stone pathways down to the village, past the thatched roof house near St. Michael's church, past the King's Head pub where the sounds of pint glasses clinking and smoke drifted out the open windows, and past the tiny front gardens overflowing with pastel-colored flowers. Sutton Bonington (population 1,600) was a long, narrow village filled with ancient buildings reminiscent of a postcard you would send home with "Greetings from England!" written on the back. Beyond the newsagent and post office was Pasture Lane, a small grocery store that made "cobs" (sandwiches). Larry and I would order two cobs and a caramel square for dessert and sit on a bench eating our lunch in the warm sunshine, all the while chatting about my next assay. Then it was back to the lab.

I worked hard and turned out to be a great lab tech, so the university officially hired me. Once I had completed Larry's projects, another researcher employed me as his lab assistant for a few months. Meanwhile, Professor Lamming was "retiring" as head of the department. The university retained him as professor emeritus, but his office had to be moved to an adjacent building. His current office was a disaster. I had extra time on my hands, so I offered to help him set up his new workplace. It took me three weeks to move and organize decades' worth of material. One day I approached Prof and told him I wanted to be his secretary.

He laughed, "I don't need a secretary."

"Yes you do," I responded bluntly, "and I would like to be it."

"We'll see," he said, smiling.

Two days passed before Prof stopped me in the hallway. "I talked to administration," he said, "and you have a job." I was so happy, I hugged him.

We spent the first month or so adjusting to each other. Prof was accustomed to secretaries who were content to get his coffee, answer the phone, and type letters. I wanted to be more than a secretary; I wanted to learn. When Prof dictated letters, I questioned statements he made and asked him to explain things I didn't understand. Science was in my blood, and Prof was involved in so many different intriguing projects: the reproduction of sheep and cattle, maternal recognition of pregnancy, interferon, and—what fascinated me most—BSE.

At the time, Britain was in the throes of the "mad cow," or BSE (bovine spongiform encephalopathy), crisis. BSE, a neurological disease that affects cattle, was first discovered in southern England in 1985. By 1990 thousands of cases were being diagnosed across the UK each month. British cattle exports were banned, beef prices plummeted, and farmers struggled not only with the loss of markets but with paltry compensation for animals slaughtered by the government.

Prof was head of the UK Animal Feedingstuffs Committee and a member of the European Union Scientific Veterinary Committee (SVC). These two committees advised the British and EU governments about matters relating to TSEs (transmissible spongiform encephalopathies), the class of diseases that includes BSE in cattle, scrapie in sheep, Creutzfeldt-Jakob disease (CJD) in humans, chronic wasting disease (CWD) in deer, and others.

Scrapie has been known about for hundreds of years, CJD was first diagnosed in the early 1900s, and CWD was discovered in Fort Collins, Colorado, in the late 1960s. So BSE was the "new kid" on the block. Because Prof had worked with sheep most of his professional life, he first taught me about scrapie. Scrapie received its nomenclature from one of its clinical signs: through excessive scraping and scratching the animal would lose its wool in large pieces. I learned that there are different strains of scrapie, that DNA testing can show genotypes of animals that are susceptible or resistant to scrapie, and that more than 90 percent of all cases occur in Suffolk and other black-faced sheep breeds.

Sheep flocks could be monitored for scrapie. If a flock was without symptoms and closed (no new female animals join the flock) for five years, the flock was considered scrapie-free. Curious to learn more about this ancient disease, I went through Prof's files and read every detail surrounding scrapie and the scrapie surveillance program. Prof was responsible for establishing the first government system to monitor and prevent scrapie. This system became European Union Directive 91/68.

Next I researched the feeding of meat and bonemeal (MBM) to ruminants, because this practice was the suspected cause of BSE at the time. The rendering industry gained tremendous markets from the sale of glycerine (used to manufacture explosives) during the two world wars. Once the Second World War was over, the rendering industry's focus shifted to further industrialization of rendering slaughterhouse waste, specifically using protein by-products from the animal-rendering industry in animal feed to improve growth rates and productivity of livestock operations. University research showed that increased protein in an animal's diet resulted in greater productivity—

dairy cows produced more milk and beef animals more muscle. Consequently, rendering companies created "meat and bonemeal" from the inedible remains of slaughtered animals, a product high in protein. Once this additive was tested on animals, studies confirmed that productivity improved, and it was marketed to farmers. The result? Forced cannibalism for vegetarian animals. Any concerns about the health effects of feeding animals back to herbivores were probably quickly silenced. "Just think how fast your cows will grow! The faster to market, the more money in your pocket!"

In the mid-1970s there were changes in the rendering process in the UK due to the oil crisis. To conserve fuel, the greaves (precursors to meat and bonemeal) were not heated to the temperatures previously required, and the use of solvents that helped extract a higher percentage of the tallow was abandoned. Scientists theorized that the lower temperatures and lack of solvents allowed the BSE infectious agent to survive and spread through feed.

"But meat and bonemeal have been fed to ruminants around the world," I told Prof one day. "Why aren't cases of BSE springing up everywhere?"

No one, not even Prof, had an answer to this question.

"There has to be another cause," I thought. The meat and bonemeal may have exacerbated the problem, but why were hundreds of cows contracting the disease weekly in Britain while only a few cases had ever been diagnosed in the rest of the world?

"Did you know there is a disease called 'kuru' found in the tribespeople of Papua New Guinea?" Larry asked, as we sat in my office reviewing papers for Prof's next Feedingstuffs meeting.

"What is it?"

"Kuru seems similar to CJD. Scientists think the natives contract it through their cannibalistic traditions. The men hunt while the women do the gathering and gardening. The sexes tend to live separately, which means the women have less protein in their diet. Women are also responsible for the burial procedure, and in their traditions they often eat the remains of deceased loved ones. In fact, the closest female relative of the deceased is given the brain, which is considered an honor. The tribespeople believe they will acquire the knowledge of the loved one who passed away."

I admire Larry for his incredible intelligence, but I told him I had no desire to eat his brain if he died first. Prof walked in as we were laughing, and I recounted the story.

"Let's find out more about that 'kuru,'" Prof encouraged me. I searched the computer archives and discovered articles by and about Dr. Carleton Gajdusek. Gajdusek lived with the indigenous people of Papua New Guinea, became their friend, and studied their way of life, including their eating habits. He discovered that kuru affected mostly women and some children. It was nicknamed "the laughing disease" because near the end of the always-fatal disease's progression, the victim had extended periods of uncontrollable laughter.

Gajdusek had found an interesting connection: the brain pathology of kuru was very similar to that of scrapie.

"What about people who eat sheep brains?" Larry asked. "Is there an increase in CJD in those populations?" The family was gathered around our dining room table eating lamb chops, curried rice, and broccoli. The children looked up briefly from their dinner plates but then returned to their meal. They were unfazed by the conversation, now familiar with Larry and me talking about TSEs in graphic detail.

"Are there any regions with a high incidence of scrapie where people still eat sheep brains?" I asked.

"Well, I know the French do, and I think people in Iceland," he replied.

A little research confirmed he was right. Icelanders have eaten sheep brains for as long as they have raised sheep, their sheep flocks have carried scrapie for hundreds of years, and Icelandic shepherds are known to consume infected sheep. Yet statistics show that CJD in Iceland occurs at an even lower than "normal" rate. (Scientists do not know why some people will unexpectedly contract CJD. This is referred to as "sporadic" CJD [sCJD]. Other forms can be traced genetically. The incidence of sCJD is estimated at one in a million worldwide.)

Because humans could eat infected sheep brains and not come down with the disease, did this mean there was a species barrier? Some scientists suggested that cattle contracted BSE by eating scrapie-contaminated feed. Yet Britain had scrapie for hundreds of years, and cattle grazed side-by-side with scrapie-infected sheep without evidence of a bovine TSE. Wasn't it more likely that the cows were eating feed contaminated by a cow suffering from BSE?

And what about the meat and bonemeal produced in Britain? Where did it all go? Was it only fed to British animals? The British government, concerned that meat and bonemeal might be the mode of transmission for the infectious BSE agent, had banned the feeding of meat and bonemeal to ruminants (cows, goats, and sheep) in 1989. This left renderers with stockpiles.

Meanwhile, the United States did not even consider BSE a potential problem. Because BSE was thought to be an anomaly, a disease of the United Kingdom and a result of the change in their rendering practices, there was no ban on feeding meat and bonemeal to ruminants in North America. In fact, according to United States Department of Agriculture (USDA) statistics, feeding meat and bonemeal to dairy cattle in the U.S. reached its highest level in 1989 and 1990. For the British renderers, the United States and Canada made great dumping grounds.

I was fascinated by the world of TSEs, but when our positions at the University of Nottingham came to an end in 1993, I thought my studies of TSEs would become an interesting hobby, a small footnote in this chapter of my life. Never did I imagine that it would mark the beginning of a personal battle steeped in the science and politics of this disease.

2

VERMONT DREAMS

England made us brave. After three years of working and traveling around Europe with the children, making the decision to move to Vermont and start our own business seemed easy. Britain was wonderful, and Larry received job offers at universities in Scotland and England, but he was ready to listen to his heart and pursue a life and work of his own design. Plus we missed our families and were eager to have the children spend more time with their grandparents and cousins. With $7,000 in savings and plenty of great ideas for ways to make a living, we felt ready for our next adventure. Larry and I loved being around each other, inspired each other, made a great team, and wanted to raise the children on a farm. Our dream was to have an agricultural business that would involve the whole family and be financially viable.

By September 1993 Larry's contract ended, but I was still working with Prof as assistant editor on *Marshall's Physiology of Reproduction*, a college reference book. A few months earlier, my parents had moved to Vermont and offered that we stay with them while we looked for a place of our own. So Larry and the children went to Vermont, and five weeks later I flew into Boston to join them.

On the plane I reminisced about the close friends we had made in England. The night before Larry and the children set off for Vermont, our friends threw a huge "going across the pond" surprise party and showered us with gifts and love. Even Prof came to hug us good-bye and wish us luck.

"What would Vermont be like?" I wondered, gazing at the clouds out the window. I had visited Vermont only once for two short days. Most of my childhood was spent in southwestern New York, in a rural setting where people kept to themselves. Our nearest neighbors were more than a quarter of a mile away, and we barely knew them let alone socialized with them. "Would the people in Vermont be as friendly as those in England? Did they take the time to get to know their neighbors?" Life in Sutton Bonington moved at a slow, enjoyable pace. People always made time to stop for a "cuppa" and a chat.

As the plane began its descent I could see the skyline of Boston through the small window. "Did we make the right decision?" I wondered, staring at the skyscrapers and the sunlight reflecting off them. The ocean sparkled below, and I could discern the outlines of boats. Just a few short weeks ago Larry and I were excitedly making plans for our new life. Now my heart ached for England as I thought about our beautiful home and garden and the security of the life we were leaving behind. All of a sudden I didn't feel so brave.

"Wake up, Hon. Wake up." Larry tried his best to whisper but was too excited.

"What is it?" I asked, rubbing my eyes.

"Look at the view!"

Larry was propped up on his elbow and motioned toward a nearby window. It was dark by the time we had arrived in Vermont the night before, and, exhausted from traveling and happy to be back with Larry and the children, I had quickly fallen asleep.

I rolled over and sat up. After spending the past three years in the flat midlands of England, it was startling to see the forested Green Mountains to the west. The scene was breathtaking. The morning sun cast a shade of purple on the ridges and a light dusting of snow covered the ski trails as they raced down the hillside. The trees were a kaleidoscope of color, changing hues with each new ray of the rising sun. The sky was crystal blue, the air clearer than I had seen in years. I turned and looked at Larry, realizing that wherever he was, I was home.

Every good scientist starts a project by doing extensive research, so as soon as I unpacked, Larry and I went to the University of Vermont (UVM) library. We gathered books and magazines about anything related to agriculture that interested us, read through them, and ran our ideas past each other. We talked about everything from having a B&B with a demonstration farm to raising rare breeds of animals. But Larry kept coming back to the idea of milking sheep. While studying in the campus library at Sutton Bonington one night, Larry had found a *British Sheep Dairy Association* newsletter and had become obsessed with the idea of raising dairy sheep. "That's what we will do," he had told me excitedly. "We will milk sheep."

"Milk *what*?!" I exclaimed. But Larry just smiled.

In 1993, there were approximately twenty-five sheep dairy farms in the United States milking nearly 3,000 ewes, with each ewe averaging only 100 pounds of milk per year. For some unknown reason, dairy sheep were never brought into the United States, only meat and wool breeds. Therefore, American farms were at a distinct disadvantage compared to farms in Europe, which were milking sheep that produced an order of magnitude more milk compared to their American counterparts. The annual total of U.S. sheep's milk production was 300,000 pounds, which resulted in 60,000 pounds of sheep's milk cheese. One day while in the UVM library, Larry discovered the Department of Commerce import figures for sheep's milk products. In 1992 more than forty-eight million pounds of sheep's milk cheese was imported into the U.S. from Europe. Therefore, less than 0.1 percent of that amount was being produced domestically.

Over dinner with my parents that evening we discussed our findings. There was incredible promise for a dairy sheep industry in the United States, but the missing link was productive dairy sheep. With Larry's background in genetics and reproductive physiology, and our passion for Europe and traveling, the idea of importing dairy sheep (or at least semen and embryos) seemed a perfect fit. The next day we began our business plan.

It was full steam ahead the following Monday as Larry phoned the Vermont Department of Agriculture to find out how to get dairy sheep from Europe. Each state has a veterinarian from the USDA working within their state agricultural agency whose job it is to monitor national and USDA programs. Dr. Wayne Zeilenga was the USDA's veterinarian for Vermont. Wayne explained to Larry that there were not yet any specific import protocols worked out by the USDA. He recommended that Larry contact Dr. Roger Perkins at the USDA's Animal Plant and Health Inspection Service (APHIS) in Maryland.

Dr. Perkins had a strong southern accent, spoke slowly, and you could sense the smile on his face as he talked. Nearing retirement age, he looked forward to completing his career with the USDA. "Y'all need to contact countries that have dairy sheep and then protocols will have to be worked out for importing from each individual country," he said. He also mentioned that the regulations for importing sheep were about to be updated. No longer would the animals have to spend five years in a federally approved quarantine (located at a USDA facility or a USDA-approved university); the new regulations would require only two months quarantine in the country of origin and one month quarantine in the United States. Both quarantines would be USDA-approved, and a series of identical health tests would be run on the animals in each quarantine. The reason for changing the regulations was that the diseases of concern had either been eradicated or could now be effectively detected by lab tests.

Armed with this information, we researched countries with dairy sheep populations and contacted embassies around the world. The embassies put us in touch with their local universities or organizations involved with milking sheep, and we discovered that the best dairy sheep breeds were the East Friesian, Lacaune, and Awassi.

Once we narrowed our list of possible source countries to England, Germany, Austria, Belgium, the Netherlands, New Zealand, and France, we again contacted Dr. Perkins. "We have an excellent rapport with Belgium," he told Larry. "The Belgian Ministry of Agriculture has been great to work with."

It seemed simple to take his advice.

When we lived in Sutton Bonington we passed an English farm with a small sign, DISHLEY GRANGE, HOME OF ROBERT BAKEWELL, each time we drove to the market town of Loughborough for shopping. Robert Bakewell lived from 1725 to 1795 and is regarded as the father of selective livestock breeding. Although the concept of breeding "the best to the best" had long been employed by farmers, the widely held belief was that production variations among animals were largely due to differences in feeding and management. Bakewell was the first to be credited with carefully selecting animals based on production traits and pairing animals with the most desirable traits to produce superior offspring. Bakewell's improved breeds of livestock were highly sought after in his lifetime.

The more Larry and I fine-tuned the numbers for our business plan, the more we realized that the most profitable aspect of our business was not in milking the sheep and making cheese but in selling dairy sheep breeding stock. To obtain the breeding stock we had three options: 1) import semen, inseminate American ewes, and through years of selection develop a high-quality American dairy sheep; 2) import purebred dairy sheep embryos, implant them into American ewes, and have the resulting offspring be the basis of a purebred dairy flock; or 3) import purebred animals.

After many, many phone calls with the USDA it appeared option 1 was the easiest, with options 2 and 3 being progressively more difficult. Dr. Perkins told Larry the new regulations would allow importation of germplasm (semen and embryos) into the United States from countries free of foot-and-mouth, rinderpest, and brucellosis. The imported germplasm could be used only in flocks that were enrolled in the U.S. scrapie certification program. Still feeling confident, we chose option 2 and based our business plan on the ability to import embryos.

The quest for a farm was taking longer than we could have imagined. When we moved in with my mother and father I warned Larry, "I love my parents, but three months is all." Little did I know that our three-month "visit" would turn into three *years*. Luckily we had a very close relationship with them. Father to five daughters, my dad enjoyed having a man in the house, a kindred spirit. And the fact that Larry is a wonderful cook was an added bonus. My parents were extremely generous, allowing us to live with them rent-free. But we still needed a source of income while creating our business. So Larry worked as a substitute teacher at the local schools, I waitressed the breakfast shift at a restaurant near one of the ski resorts, and together we established a freelance editorial services business.

Because our bedroom was also our office, and the children's bedroom could only be accessed by going through our room, we often had difficulty concentrating as the children ran back and forth. One day after the third interruption within a fifteen-minute period, I threw up my hands in exasperation. "I can't do this," I complained to Larry. "Editing takes my full concentration, and with so many interruptions I have to go over and over the same page. I've been on this same page for the last ten minutes!"

"I'll take care of it," Larry said, jumping up from his chair and storming out of the room.

I sighed and put my head on my desk. I was tired of being tired. Larry and I worked late into the night on editing and the business plan. He would drive me to the restaurant every morning, so we were both up by 4:30 A.M.

I didn't want him to yell at the children. They were just being kids. But I knew that when I was upset, Larry would do everything he could to settle things.

I listened. No yelling and screaming children. No yelling or screaming father. Just silence.

A few minutes later, Larry walked in smiling and shut the door behind him. "You should be able to edit now," he said, bending over to kiss me on my head.

"What did you do?" I asked nervously.

"Don't worry about it. Let's just get our work done," he answered, picking up his copy of the business plan and sitting back down in the recliner. For the next hour the only sounds I heard were of the children playing outside and birds singing in the trees. I was so engrossed in editing that I jumped when there was a soft tap on the bedroom door.

The door opened quietly, just enough for Heather to stick her head in.

"I'd like to use one of my passes," she said, looking toward Larry. When he smiled and nodded, she quickly hurried toward her room, stopping briefly to hand me a piece of paper.

I unfolded the crumpled piece of lined paper and read in Larry's handwriting: *This entitles the bearer to one quick trip to their bedroom.*

I smiled. "What is this all about?" I asked him.

"The children have three of these they can use each day," he said with a grin. "That will give us some more peace and quiet."

Heather quickly and quietly ran back out the door, a plastic grocery bag stuffed to its full capacity with an assortment of doll clothes and cooking supplies under one arm and a small log dressed in a baby outfit under the other.

"Thank you," she said softly.

"I won't ask," I thought, looking at the pine log and shaking my head. "I just won't ask."

We attended any meeting that might even distantly relate to our project: the Vermont Sheep Breeders Association (VSBA) annual meetings, the Vermont Sheep Summit, the "Future of Vermont Agriculture," Small Business Administration meetings, the "Vermont Food & Agricultural Products Export Workshop" sponsored by Senator Patrick Leahy, and even meetings with the Dairy Goat Promotion Board.

Senator Leahy was involved in a number of initiatives that impressed us both. When we observed how he fought against the use of bovine growth hormone, otherwise known as BST, in dairy cattle, and advocated for agriculture in general, we decided we should meet with him and his staff to talk about our project. So in early December 1994 we went to Montpelier, Vermont's state capital, about a half-hour drive from my parents' house.

After going through metal detectors, we rode an elevator to the third floor of the federal building. At the end of the hallway was a door flanked with the American flag and the Vermont flag. "I guess that's Leahy's office," Larry said, squeezing my hand. We were both nervous. This would be the first time either of us had personally met a senator. As we entered we were greeted by a secretary, and a few minutes later we were ushered into a conference room. Bob Paquin, Senator Leahy's chief aide, rose from the large conference table, shook our hands, and invited us to sit. Tom Cosgrove, another aide, joined us by speakerphone. "Thanks for coming in," Bob began. "The senator has put us in charge of helping to diversify Vermont agriculture."

Tom and Bob responded enthusiastically when Larry and I shared our business plan and our ideas for the incredible potential for sheep dairying not only in Vermont but throughout the United States. Tom inquired about the import regulations, and we told him about our many conversations with Roger Perkins over the past year and the various countries we considered working with.

"Do you know what the opinions of the VSBA and the American Sheep Industry Association are concerning your importing?" he asked.

I smiled as I thought about our first VSBA meeting. It was exciting to be around sheep farmers, listening to them discuss their flocks, watching as they knitted with the wool, talking about what we were still dreaming of.

My daydreams were interrupted as Larry told Tom about the positive feedback we received from VSBA members. "And as far as ASI [the American Sheep Industry Association] is concerned," he continued, "all VSBA members are automatically ASI members. While we haven't had any direct contact with anyone in the national organization, we don't foresee any problems. There's tremendous promise for the sheep dairy industry. The U.S. sheep population is declining, the price of wool is falling, and any stimulation would be positive." Tom and Bob agreed.

As the meeting came to an end Tom said, "I'll be glad to call APHIS to check up on the status of the regulation change."

Bob shook our hands. "Great to have folks like you come in."

True to his word, Tom Cosgrove called the USDA on our behalf. At the annual VSBA meeting the following month, we spoke to Phil Hobbie about our conversation with Tom and our long wait for the regulation change. Phil was VSBA's representative to ASI and a Vermont sheep farmer. He told us he too would "put a word in" for the regulation change at the upcoming ASI meeting in Washington, D.C. He also expressed his interest in obtaining semen from some European meat breeds of sheep.

One of the best parts of planning our business was our family discussion around the dinner table each evening. When asked about how they would like to be involved, Francis expressed interest in pasture and sheep management and longed for a dog, Heather was fascinated with the idea of milking the sheep and wanted a llama to guard them, while Jackie was determined to help take care of the sheep as well as make cheese from their milk—high aspirations for children now ages nine, seven, and six years old. Ours would be a true small family-owned and family-operated business.

In March 1995 we set off for the Great Lakes Dairy Sheep Symposium and the Wisconsin Sheep Breeder's Conference in Madison, Wisconsin. This would be the first sheep gathering the entire family attended. Before we had driven half an hour, the bickering started in the backseat. Francis was winding up his sisters. Jackie wanted to sleep, and he kept talking. When I turned around, Heather seized her opportunity. "He's taking up too much room! He's on my side!" she whined. Larry and I sighed and looked at each other in frustration. Our small car was loaded to the brim, and the three-day trip ahead of us seemed daunting. It was time to be creative. I told the children, "Each of you now has ten dollars. Every time you fight with each other, you will have to pay Daddy and me a dollar. Whatever you have left by the time we get to Wisconsin is yours to keep." It was the best twenty-seven dollars we ever spent.

When we arrived in Madison the children were excited and a little nervous to be attending talks on their areas of interest. In the morning we would schedule the day, and throughout the weekend we would reconvene in our hotel room to review the talks. I will never forget the image of Heather, tiny for her age, sitting on the edge of the bed with her legs dangling over the side, talking about the treatment and pre-vention of mastitis in ewes. Francis made many friends and through various workshops learned to handle sheep, as well as hoof trimming and the basics of shearing. Jackie, quiet as usual, had studiously taken extensive notes, all of which were neatly written in her notebook. On Saturday evening there was a large banquet featuring sheep's milk cheeses, and Jackie wandered from buffet table to buffet table, eyes wide with admiration, sampling the different varieties. Another high-light for Heather and Jackie was meeting Olivia Mills. Olivia was the author of the only book published in English about milking sheep, *Practical Sheep Dairying*, and was responsible for the resurgence of sheep dairying in Britain.

The focal point for Larry and me was meeting Jock and Hilary Allison of Silverstream in New Zealand. A very nice couple in their midfifties, they imported purebred East Friesian sheep from Europe, collected embryos from them, implanted the embryos into New Zealand ewes, passed a very stringent five-year quarantine, and were now selling East Friesian embryos throughout the world. New Zealand was on the USDA list of approved countries for importation. After meeting the Allisons, we felt one step closer to our goal.

Vermont had established a name for itself in sheep importation in the early 1800s. As a result of Thomas Jefferson's Embargo Act of 1807, the shipping battles between Britain and France, and the War of 1812, the United States lost much of its international trading business and was forced to rely on its own resources for the supply of wool prod-ucts. From 1809 to 1810 William Jarvis, the U.S. consul to Portugal and a native Vermonter, imported into his home state the first Merino sheep from Spain and Portugal. Merinos are an exceptional wool breed, and by 1837 Vermont had more than one million sheep grazing its hills. The Civil War caused the price of wool to jump from 25 cents a pound to $1.50 per pound, further increasing the popularity and the purchase price of Merinos. Vermont Merino rams sold for $500 to $3,000 with some individuals fetching more than $10,000—an enormous sum in that era.

But with the railroad expansion and farmers' access to the grasslands and vast pastures of the west, Vermont sheep farmers found it difficult to compete. Sheep could be raised in the west for as little as 25 cents a head, while it cost more than a dollar on the East Coast. Vermont farmers discovered dairy cattle to be more profitable, and sheep were soon replaced. By the 1950s Vermont had more than ten thousand dairy farms.

Two world wars brought more change and the advent of the second industrial agricultural revolution, but this time Vermont agriculture did not benefit from the conflicts. Once the wars ended, the manufacturers of explosives did not want to close their plants and layoff their employees, so, together with scientists, they found a way to market products and by-products of their industry to the agricultural world: Hence the creation of chemical fertilizers, insecticides, and herbicides. Along with this was the shift from horses to tractors. Vermont's small hill farms were not suited for large machinery or large-scale production, and they struggled to survive as the average-size farm in the United States grew from 145 acres in 1925 to more than 450 acres by the mid-1990s. By 1994, Vermont was 80 percent wooded and only 2,056 dairy farms remained. The loss of farms in the state continues at a rate of four farms per month.

But agriculture and the rural landscape were vital to Vermont's economy, and many organizations made it their mission to support farmers. One of the people most aware of the loss of farms and the changing face of Vermont's rural landscape was Vermont philanthropist Houghton Freeman. He and his wife, Doreen, devoted their lives to preserving and enhancing life in rural Vermont and fostering relations between the United States and Asia. In the 1940s, after saving some money while in the Navy, Mr. Freeman purchased a farm in Greensboro, Vermont. His father fell in love with the beautiful location and moved there to raise cattle until he passed away in 1992. Houghton Freeman was very close to his father and, in honor of his memory, wanted to run a successful agricultural operation that could potentially be a model to other Vermont farmers and help revitalize agriculture in the state. He researched many endeavors including beef cattle, emus, and llamas. Then he came across our business plan, was intrigued, and arranged to meet us.

Happy, nervous, excited—all these feelings seemed to happen simultaneously as Larry and I donned our best suits and headed north to Stowe, Vermont. The Freeman's offices were in a fairly nondescript, small shopping center. Andrea, their secretary, greeted us at the door and took our coats. The office was beautiful with pale yellow walls, cherry furniture, and a large, comfy, overstuffed couch. The walls were covered with paintings of rural landscapes and photos of projects the Freemans and their family foundation supported.

Mr. Freeman welcomed us, smiling as he eyed our suits. He was dressed in slacks, a well-tailored sweater, and leather slippers. With natural dark hair and twinkling eyes, his handsome looks had not faded over his seventy-plus years. Slightly shorter than her husband, Mrs. Freeman held herself with grace, an understated elegance. She walked over, shook our hands, and, in a soft English-accented voice, asked if we would like a cup of tea. Instantly I felt a close connection.

For an hour we talked about our dreams of having a sheep dairy. It was obvious from their questions that Mr. and Mrs. Freeman had researched the idea as well. The only stumbling block was how and when we would get the genetics. Larry chronicled his conversations with the USDA, how we had been waiting for more than a year for the regulation change, and our hopes for being able to import embryos in the very near future. Mr. Freeman gave us a list of queries and said he was intrigued by the idea of having dairy sheep at his Greensboro farm. We assured him we would answer his questions as soon as possible and would keep him posted when the regulation change went through.

3

EUROPEAN SHEEP IN THE
GREEN MOUNTAINS

There were days when I questioned whether the sheep project would really work. By the spring of 1995 we were using some of our savings to pay our bills, and every time we spoke with Roger Perkins to get an update on the regulation change it was always the same response: "It's limping through the printers."

And then something in the heavens shifted. On a sunny spring morning we received a letter in the mail notifying us that we had been awarded a $4,500 grant from the Connecticut River Valley Partnership Program to travel to Europe and research sources for dairy sheep genetics. Larry and I were overjoyed. Now we could actually visit the farms, see East Friesians, and our dream would be one more step closer to reality. Our first stop was Britain.

We spent the night in a quaint English inn, and by 6:00 A.M. the rooster in the yard was putting on a full show, entertaining his harem of hens while a brown and white rabbit stretched in the grass and bathed in the early morning sun outside our window. After a proper fry-up breakfast, Larry and I headed to a pretty village near Stratford-upon-Avon. We knew we were at the right place when we spied the SHEEP'S MILK 75P sign.

A young woman walked us to the barn to meet her father who had a ewe propped up against his legs, hoof trimmers in hand. He explained it was a very wet spring, and his sheep had foot problems.

"How many sheep do you have?" I asked.

"Oh, somewhere around four hundred or five hundred," he said, lifting up the back leg of a particularly large ewe. "We have some purebred East Friesians, but mostly crosses with either Dorset or Texel."

"Do you mind if I film?" I asked, pointing to the video camera I carried.

"Not at all, feel free," he answered happily.

"How often do you milk?" Larry asked.

"My daughter handles the milking," he said, motioning toward her, "and the record keeping."

"I'll show you the milking parlor while my dad finishes up," she said. As we walked toward the milk house she explained that they had 110 acres and four people working the farm, including her husband and a farmhand. "Today I'm milking 306 ewes," she mentioned as we entered the spotless milking parlor. "We milk year-round and sell most of the fluid milk to a place in North Yorkshire. My dad delivers three times a week. It takes him all day to make the trip."

"Would you like to see more of the girls?" her father asked, popping his head around the corner. We followed him to the open barnyard, and when he whistled commands to his dog, the black and white Border collie rounded up a large group of ewes.

Larry and I stepped to one side of the pen, while the farmer stood on the other.

As I zoomed in with the camera, I scanned the group. "Which ones are the East Friesians?" I quietly asked Larry.

"See the ones with the big ears, a bit taller than the rest?"

"You mean the ones in the front of the group, heading straight toward us?"

"Yep, that's them."

"Where did you get your East Friesians?" Larry asked.

"From the Netherlands, a woman by the name of Kostense. I bought some rams from her," he yelled as the sheep charged past, ignoring us and searching for food.

The next farm milked 210 ewes. The problem here was the couple did not keep records and spent almost six hours milking. Again, they had some purebred East Friesians, but mostly East Friesians crossed with Texel. They made wonderful ice cream from the sheep's milk, but did not market the fact that it was sheep's milk for fear it would "put people off."

It was not what I had expected. In their milking parlor I watched twelve ewes enter the stanchions to be milked, while the remaining ewes in the barn climbed on top of each other, trying to be the next ones in. "What are we getting into?" I wondered. There was no way I wanted a life like this. The poor couple looked frazzled and exhausted.

On the plane ride from England to Brussels the reality of our situation set in, and I felt overwhelmed. Eighteen months earlier we had given up the security of a weekly paycheck to pursue making a living in the country. Our business plan and its success were based on acquiring the best available dairy sheep genetics and starting an industry, but a viable industry could not be accomplished with the animals we saw.

Perceptive as always, Larry saw my face and said, "You have to understand, Hon, the last shepherds were milking for six hours every day because they did not keep records. They don't know who their best producers are, therefore when they breed they are not selecting for milk production. And both of the farms we visited were milking mostly crossbred sheep, not purebred East Friesians."

"I know, I know," I said. "I really wanted to work with farmers in England, but if those two farms are any representation . . ." I looked out the window.

"Let's not give up hope yet," Larry said, turning my chin toward him, forcing me to look him in the eyes.

I looked down, trying to avoid his gaze. We both worked hard on the project, but there was no doubt he had put the most effort in. He had tremendous patience waiting for the regulation change, convinced that sooner rather than later it would go through.

"Look at me," he said softly, and I looked up. "What are you thinking?" he asked.

"About the business plan, about our future," I said. Here we were in Europe searching for the best sheep and we didn't even have a farm of our own yet. We spent all our free time scouring the state from top to bottom looking for a farm. We were on a first name basis with real estate agents in every county of Vermont. We looked at small farms, large farms, farms in the mountains, farms in the valleys, farms with beautiful houses, farms with barns more beautiful than the houses, farms without barns, and farms without houses. But our hearts were in the Mad River Valley where my parents lived. The very first real estate agent we spoke to, Anna Whiteside, was from the valley and turned out to be our favorite. She tried to find something suitable nearby, but the valley had lost many of its farms to tourism and second-home owners and land prices had gone up as a result.

So we kept looking, and looking. The children complained of spending so much time riding around the state. They wanted to be home playing with their friends. "Why can't we find a place in Warren?" they groaned.

The plane hit a small amount of turbulence, and I was jolted back to reality. A few months prior, the Belgian Embassy had put us in contact with their sheep expert, Dr. Bernard Carton. Soon we would meet him, but would we be able to work with him? When Larry and I had faxes flying back and forth to arrange our visit, Dr. Carton kept insisting, "You *must* see the Beltex [a breed of meat sheep]." We would respond "We want to see East Friesians." To which he would counter, "You *must* see the Beltex." Finally we said, "*If* we see the East Friesians, we will see the Beltex."

As the flight crew prepared for landing, Larry leaned over and hugged me. "Trust me," he said. "It's all going to work out. All of our research showed Belgium and the Netherlands to be the best source for the best animals."

I envisioned Dr. Carton to be a small, slight-framed man in his late fifties with very serious features. When he arrived at 9:00 the following morning to drive us to the farms I discovered I was wrong on all accounts. Standing slightly over six feet tall, Dr. Carton was in his early forties, well-built, with a boyish smile.

Traveling throughout Flanders and the Netherlands over the next few days gave all of us ample time to get to know each other. "Carton," as we soon referred to him, studied veterinary medicine in Belgium and the United Kingdom. His passions were sheep, politics, his wonderful family, and great food. With the exception of politics, we shared the same passions.

Carton readily answered our many questions and asked quite a few himself, trying to figure out what these two Americans were all about. We told him of our business plan and the disappointments in Britain. As we spoke, his smile grew. He knew we would like what he had to show us: purebred sheep flock after sheep flock, beautiful farms with barns where you could practically eat off the floor, and incredible records. The milk production records went back more than a decade and showed ewes averaging 1,000 pounds per lactation. One line of sheep averaged more than 2,500 pounds! The shepherds also had flock records and individual conformation records. Sheep dairying was a serious business here.

Sheep have been milked for two thousand years longer than cattle, and more sheep are milked worldwide than cows. Hundreds of years ago, the East Friesian breed originated from sheep that came from India and made their way to northern Germany and Friesland in the Netherlands where they thrive today. (Friesland is also home to the Holstein Friesian breed of dairy cattle.) The East Friesian sheep is of different genetic lineage than many of its European counterparts, as evidenced by its long hairless "rat" tail. It also has large ears, long legs, and sheds the wool on its belly during the summer. The breed is very personable, not nervous or jumpy, and will go to any measure for food.

The shepherds matched the breed; they were warm and welcoming. Every visit to a farm started with a cup of coffee—strong European coffee—in their house. Larry and I didn't drink coffee but, not wanting to be rude, would graciously accept a cup. When everyone finished, we would walk to the fields or barns to see the animals, review records, and videotape the flocks and facilities. Then it was back to the house for another cup of coffee before we left for the next farm, where the process would be repeated. In the course of a day we would visit five or six sheep farms. At night Larry and I lay in bed, unable to sleep, buzzing from caffeine.

And then we finally saw the Beltex.

It was love at first sight for me. The Beltex look like pigs with wool and are very gregarious. While East Friesians are tall, graceful, and elegant, Beltex are the opposite: incredibly compact and muscled. Crossing a Beltex ram with a ewe of any breed produces much meatier lambs; a trait that Larry identified immediately and knew could have enormous value in the American lamb industry.

I was smitten with their personalities. Every time we walked into the field they gathered around my legs like chubby little puppies, looking up happily at the camera as I filmed.

Carton was right: We had to have the Beltex.

The evenings were ours to explore Ghent. After Carton dropped us back at the hotel, we would make our way to "Stadcentrum." The center of Ghent is similar to Brugge but without the throngs of tourists. Beautiful, elaborately carved stone buildings line the canals, while scattered throughout the city are plazas with umbrella-covered tables ideal for lazy afternoons spent people-watching or for sharing an evening drink with a loved one in the warm night air.

And the restaurants! Incredible fresh food: wine, cheeses, seafood, meats, and vegetables in beautiful preparations. The Belgians produce more than three hundred styles of cheese, craft more than three hundred beers (each served with its own unique glass), and consume more champagne than the French. This was a culture after my own heart.

One morning we traveled an hour east of Ghent to the city of Aalst, host of "Dag van het Schaap." The "Day of the Sheep" is the largest sheep-related festival in Belgium, and in 1995 more than five thousand people attended. The day flew by as we looked at the various sheep breeds, many of which I had never seen: short sheep, tall sheep, thin sheep, fat sheep, sheep with black faces, sheep with white faces, and sheep with red faces. There were many breeds I had never heard of: Zwartbles, Rouge de l'Ouest, Clun Forest, and a blue-tinted breed called Bleu du Maine. The festival was an excellent opportunity for us to meet with various sheep breeders, university researchers, and government officials.

Before our trip Larry had the foresight to arrange a meeting with officials from the USDA and the Belgian Ministry of Agriculture in Brussels to discuss importing/exporting sheep germplasm. Danielle Borremans was a Belgian working for the American embassy and was as friendly in person as she seemed from her correspondence. It was a gorgeous mid-June day when we met her at the American embassy and walked together to the Belgian Ministry of Agriculture where we were greeted by Dr. Mark Dulin of the USDA and Dr. Maillet of Belgium. As we gathered in the crowded quarters of Dr. Maillet's office, we explained our business plan.

"We are interested in importing East Friesian and possibly Beltex semen," Larry told the group. "And maybe embryos," he added expectantly.

"What farms are you working with?" Dr. Maillet asked.

"Dr. Bernard Carton has been showing us various sheep farms," I began. "We saw East Friesians from Kostense, Tylleman, and Atema, and Beltex from Blanchaerts."

Dr. Maillet straightened proudly in her chair, a slight smile on her face. She admired Dr. Carton and was very familiar with the Belgian and Dutch farms we named. They had won many awards for their outstanding sheep. "Those are excellent farms," she said.

"We would like to import embryos if possible but are having difficulty finding a facility to handle embryo collection. Perhaps you might have some suggestions?" Larry asked.

Dr. Dulin and Dr. Maillet looked at each other. They were both silent. Then Dr. Dulin spoke up, "What about importing live animals? Would that work for you?"

My heart raced. This would put us almost five years ahead of our business plan. I looked at Larry, whose face registered disbelief.

Dr. Maillet was nodding vigorously and smiling. She was justifiably proud of her sheep industry, and this would be the first live animal importation of sheep from Belgium into the United States.

Larry looked over at me, and I couldn't help but smile, a huge smile, "That would be wonderful!"

"Great. Let's get the protocols going," Dr. Dulin replied.

It was more than coffee that kept Larry and me awake that night. We couldn't wait to get back to Vermont and share the great news with the Freemans and our families. We were already planning the importation. The animals would be lambs when they were brought over, be bred in the fall, lamb the following spring, and start milking the following summer.

As soon as we arrived home, we contacted Dr. Perkins. "Everything looks good. I'll get the paperwork going. The regulation change is moving along," he said, "and could be finalized in as short as a few weeks." Next we sent a thank-you to the Connecticut River Valley Partnership Program for the funding. We were extremely grateful. More had been accomplished in those two weeks than in the past two years.

We drove to Stowe and showed Mr. Freeman the videotapes of the sheep farms and their flock records. He was excited and asked us to help find a shepherd for his farm. Larry called different organizations and schools and found many excellent candidates, but two in particular stood out. A young couple, Frankie and Marybeth Whitten, worked at Old Chatham Sheepherding Company in New York and had extensive sheep experience: handling, management, health, pasture management, and milking. At first Mr. Freeman was unsure of hiring a couple, but we suggested that given the remoteness of his farm and the long Vermont winters, it would be better to have two people working together. Within a few weeks, Mr. Freeman had not only hired them but had also made arrangements for their housing on the farm. Larry and I were soon back to our normal work routine at home, but we could feel the momentum building.

"Can you get that?" Larry yelled to me up the stairs, as the delicious aroma of pumpkin soup drifted up from the kitchen—wonderful for the cold autumn day. I was in our office editing a manuscript and answered the phone on the third ring. "Hello, Ag-Innovations."

"Hello? Ms. Faillace?"

I immediately recognized the southern accent. "Hi, Dr. Perkins," I said happily.

"I've got some great news for you. You can start the importation."

"Really? That's incredible! Thank you!"

I ran down the stairs yelling, "It's a go! We can go ahead!"

"Go ahead with what?" Larry asked as he prepared to grate nutmeg into the soup.

"With the importation!" I said.

"Great! Would you stir the soup?" Larry asked, as he tossed the nutmeg on the counter and passed me a long-handled wooden spoon. "I have some phone calls to make."

He contacted Dr. John Hansen of the Animal Import Center in Newburgh, New York, to find out about any special requirements. Then he made arrangements for the blood tests to be done in Europe. (The U.S. tests were done by the National Veterinary Services Laboratory in Ames, Iowa.) Over the following month we were constantly on the phone and sending faxes. Roger Perkins sent us a draft of the protocol and suggested we use a customs broker to get us through the nuts and bolts of our first importation. And then Mr. Freeman called.

The snow was already falling as Andrea took our coats, scarves, and hats when we arrived at the Freemans' office. "Would you like some tea or coffee?" she asked. Normally we would say yes, but our adrenaline was working overtime and the last thing we needed was caffeine.

"No, thank you," we answered in unison.

"Glad to see you're not all dressed up this time," Mr. Freeman said, smiling as he shook our hands. I noticed he was again in his leather slippers.

As we sat on the overstuffed couch I tried to sit up straight, but each time I put my back against the couch, my legs stuck out in front of me. I shuffled to the edge of the sofa so my feet would touch the floor. "Now is *not* the time to be fidgety," I thought to myself.

"I've been going over the figures," Mr. Freeman began, referring to the various tables of projections based on milk and cheese production Larry sent him. The calculations assumed the animals would be lambs at the time of importation and not reach their full potential until their third year.

"I want to order some sheep," Mr. Freeman continued. We held our breath. This was the moment we had been waiting for. "Now, how many rams do you think I will need if I have, say, twenty-five ewes?"

"Well, it depends on the genetic diversity of your ewes," Larry replied.

"Twenty-five ewes!" I thought. "I never imagined we would sell twenty-five sheep to our first customer."

"If you have a broad enough base of genetic diversity with your ewes, you could get away with two rams. More would be better, but two would do."

Mr. Freeman sat there for a moment, thinking.

"Okay then. I want you to get me fifty ewes and two rams."

Our mouths dropped. We quickly tried to recover and act professional.

"You'll make me a deal too, right?" Mr. Freeman winked.

"Of course," Larry answered.

"Write up an agreement and fax it over to me. Keep it simple."

When the phone rang on November 29, 1995, it was Larry who answered, "Ag-Innovations."

"Dr. Faillace?"

"Hi!" Larry replied.

"This is Dr. Perkins."

Larry laughed. "I know."

"I have Dan Harpster on the speakerphone here." Dan was involved with the national Volunteer Scrapie Certification Program. Roger sounded serious. "I just wanted to let you know that the regulation change is grinding through the system," he continued.

"What do you mean, 'grinding through the system'?" Larry asked, angst creeping into his voice.

"It's just going slow," Roger replied.

"But I thought that you said we could start the importation."

"I'm sorry, but the regulation change is going slow," Roger repeated.

Dan was quiet for most of the conversation, and as much as Larry tried neither he nor Roger would give any straight answers. Finally Roger said, "It would be wrong of me to give any info that might give a competitive advantage to someone."

I was stunned when Larry recounted the conversation. But Larry was undeterred by another potential setback and contacted Dr. John Bramley, head of the UVM animal science department, about a Cooperative Service Agreement (CSA). This was an arrangement in which Larry and I would import the sheep, but the ewes would lamb and remain in a USDA-approved quarantine located at the University of Vermont for a period of up to five years, or until the regulation change went through. Dr. Bramley was very interested in working with us, and Larry's perseverance paid off a few short weeks later when Dr. Bramley and UVM approved the contract for a CSA.

In December 1995 Larry and I set off for Europe; this time to narrow down possible source flocks and take a closer look at the individual ewes that would be mothers to our lambs. Again the northern European shepherds went out of their way to make us feel welcome. One Belgian shepherd, hoping we would buy his sheep, took Larry, Carton, and me out for a sumptuous five-course lunch at a wonderful Flemish restaurant named De Schwaan. The meal began with an aperitif, followed by Belgian endive au gratin, then a steaming stainless-steel pot full of mussels in a savory broth. The main dish was steak in a peppercorn cream sauce with gently sautéed vegetables on the side. When I thought I could not eat another bite, the waiter presented a chocolate mousse, artistically designed with a profiterole on the side. The meal ended with an exquisite glass of sherry.

Satiated, we slowly made our way to the car. Carton's cell phone rang. It was Mrs. Blanchaert, the Beltex breeder we were on our way to see next. "Would the Americans like to try Belgian waffles?" she asked him.

Carton smiled his mischievous smile. Belgian waffles are much larger and fluffier than traditional American waffles and are often served with whipped cream, strawberries, and chocolate. "Yes, they would be delighted!" he laughed.

Birth announcements! From January through March of 1996 our fax machine buzzed with lambing records. Larry or I would grab the fax, get the flock records out, check the mother's milk production and conformation, and then look at her on our video. Our list of potential import candidates was expanding.

The first importation would consist of fifty-two lambs for Mr. Freeman. After calculating the cost of the animals, two quarantines, all the blood tests, the transportation, commission, and broker fees, the price we arrived at was $5,000 per animal. For Mr. Freeman we made an arrangement for a reduced price in exchange for us being able to select twenty lambs from his first two lamb crops. The twenty lambs from Mr. Freeman would be the basis for our dairy flock, and, because East Friesians are so prolific, our flock would grow quickly. Once our flock was established, we would continue to import East Friesians from Europe in order to constantly improve our genetic stock.

We sent the CSA agreement to Roger Perkins, and it was approved quickly by the USDA. When asked about the status of the regulation change Roger replied, "It was blazing through the system, but then it stumbled." So again we contacted Tom Cosgrove from Leahy's office. He offered to talk with the Office of General Counsel to try to move things along.

"Have you heard from Roger Perkins?" Larry asked. It was now mid-May 1996, and there was still no update on the regulation change.

"Not yet," I replied, a little absentmindedly as I folded some laundry and thought how much I'd rather be outside on this beautiful day. Larry walked past the fax machine and picked up a pile of faxes. Since my father also worked from home, it was common to have many faxes within a day.

"Let's take a break and go outside," Larry said, as if reading my mind.

I gratefully put down the clothes and went to the kitchen to fix a glass of my mom's wonderful homemade iced tea. I met him on the front steps.

"There's a fax here from Senator Leahy's office," Larry said as I sat down beside him.

I read over his shoulder "Federal Register . . ."

"That's not the regulation change, is it?" I asked, half jokingly.

"I don't think so."

Larry read a little more, "Wait! I think you're right. It *is* the regulation change!"

Sure enough, our three years of waiting were finally over. We read the fax together. The animals would have two months in quarantine in Europe and one month in a USDA facility in New York. If all went well, Mr. Freeman would have his sheep by August. Immediately we contacted Dr. Carton to have him start collecting the lambs. We trusted him to make the final judgment and reject any lambs we had selected that he felt might not be outstanding and to replace those with his choice. The regulation required the animals to be quarantined for two months in Europe. Because some of the animals were coming from the Netherlands we asked the USDA in Belgium if they could all be quarantined together instead of having two separate quarantines. An animal is defined as being "from Belgium" after it spends a minimum of sixty days in Belgium. The USDA said "no problem." The time in the Belgian quarantine would count.

Then we called Mr. and Mrs. Freeman to share the great news. Last, but not least, we sent thank-you letters to Senator Leahy's office and his staff for all their hard work.

As a requirement of importation, we had to enroll the sheep in the U.S. scrapie certification program. Although scrapie has been around for hundreds of years, the first diagnosed case in the United States was in 1947 in a Suffolk sheep imported from Canada into Michigan. Once the disease was discovered, a more extensive testing program was established and more cases were found throughout the United States. By the early 1990s the disease was firmly established in the U.S. Dr. Perkins suggested we call Dr. Bill Smith from the New England regional office of USDA/APHIS in Massachusetts to start the scrapie certification program in Vermont. Bill explained that despite the prevalence of scrapie in the United States, the USDA was reluctant to fund programs for an animal disease they considered of little consequence. Therefore the national scrapie program was voluntary, and most states resisted joining, including Vermont. Bill thought it was great we would institute the program in Vermont.

But our first meeting with the Vermont Scrapie Certification Board at the Vermont Department of Agriculture was anything but amicable. (There was no program, but the state had formed a board.) This was the first time the program would be initiated in Vermont, and the last thing we expected was any resistance. Wasn't this a good thing, to monitor and control a disease in sheep? Dr. Wayne Zeilenga was not so sure. He did not like the idea of sheep being imported into Vermont.

"What if they don't have proper identification?" he asked.

Larry explained that each sheep had not only tattoos but also ear tags.

"What about the risk of these animals bringing in disease?" Wayne asked.

Larry explained the series of tests the animals would have during each of the quarantines.

"What about scrapie? What if these sheep have scrapie?" he then asked.

I looked at him with incredulity, given the fact that cases of scrapie had been found in Vermont under Wayne's watch. This was the beginning of my understanding of the inconsistencies and contradictions we would face around the health and monitoring of our sheep.

"Isn't that the goal of the surveillance program, to monitor for scrapie?" I said. "Besides, none of the European flocks would be approved for export if they had a history of scrapie. All of our European source flocks were enrolled in the EU scrapie surveillance program, almost since its inception in 1992. And none of the flocks had ever had a single case of scrapie. In fact, there has never been a case of scrapie in an East Friesian—ever."

To go above and beyond what the USDA required for importation, we asked each of the European shepherds if they ever fed meat and bonemeal. Most were aghast that we would even ask; all emphatically said no. And we only imported from flocks that were enrolled in the scrapie program. This was *not* a requirement of the USDA protocol, but with our knowledge of TSEs and after working with Professor Lamming, we felt it was valuable, another reassurance for ourselves and our customers.

Dr. Sam Hutchins was the state veterinarian for Vermont, had worked for the Vermont Department of Agriculture for most of his life, and looked forward to his retirement the following year. Like Wayne, he was reluctant to start the monitoring program. The last thing he wanted was more responsibility. But despite the initial opposition, after days of discussion the board finally agreed to initiate the scrapie program. Mr. Freeman's flock would be identified on their ear tags as "VT001"—the first flock enrolled in the Vermont Scrapie Surveillance Program.

In June of 1996 we headed back to Europe for the third time in less than a year. This time Larry and I flew back from Europe on separate planes. I was on a passenger plane while Larry traveled on an El Al cargo plane with forty-nine ewe lambs and two ram lambs in hand-made wooden crates. We chose El Al for their excellent reputation, but one factor we didn't take into account was their security measures. El Al would not announce flight departure and arrival times.

The sheep had passed the USDA-approved Belgian quarantine and tests with flying colors and were extremely calm. After six hours of waiting at the Amsterdam airport, the signal was given and the place was abuzz with activity. Larry helped load the sheep onto the cargo area of the plane, and within less than an hour the pilots, crew members, Larry, and another man accompanying a horse were in the air.

I flew to New York City where my parents met me and took me to a hotel near Newburgh, New York, where I anxiously waited for Larry's call. Around 1:00 A.M. the phone rang. Larry and the sheep had arrived safely, and we hurried to the quarantine facility.

Larry hugged and kissed me. "They did great!" he said. "There were only five people on the plane. While we were in air, I was allowed to go below, down a steep flight of stairs, and check on the sheep. At the front of the cargo area was a horse in a container. He was not happy about being alone and was yelling and kicking at the door of his pen. Then I passed hundreds and hundreds of flower and vegetable crates," Larry continued excitedly. "The sheep were in the rear of the plane. At one point I had to walk sideways along the edge of the wall because there was so much cargo."

His eyes lit up like a child in a candy shop as he recounted the story. "The sheep bleated when they saw me, but mostly they just chewed their cuds. No one was stressed."

"And you should have seen the reception they got when we landed in JFK. I was standing on the tarmac with the crates of sheep and all the airport guys came running over to check them out. I bet you most of them had never seen a sheep before!"

Soon a USDA employee met us in the office and explained that the sheep were now settled. We were assured that the sheep were in good hands.

August 2, 1996, was the day we had dreamed of for three years. Larry, Francis, and I arrived in Newburgh early in the morning. The animals had successfully completed the USDA quarantine—all the tests and health inspections—and were given a certificate of inspection that stated the animals were "free of evidence of communicable disease or exposure thereto." And now they were going to their new home in Greensboro, Vermont.

Francis and I waited in the office while Larry accompanied some of the USDA employees to the building with the sheep. One woman who took care of the sheep had grown so close to them that she cried when the animals were leaving. I videotaped the animals and made sure Larry took photos of Francis standing beside the sheep trailer with the USDA sign in the background. "What a wonderful business relationship," I thought. "From Roger Perkins, to Dr. Dulin, to the Newburgh employees, everyone was extremely professional and helpful." This was the first time since 1947 that sheep had been imported from Europe into the United States and the first time ever from Belgium and the Netherlands. By working closely with the USDA we were able to make our dreams come true.

When Larry and I finally made our way home that night there was a strange fax waiting for us. One of the European breeders whom we had bought sheep from had heard a rumor that the USDA had killed the sheep. We shook our heads in disbelief, not giving it much thought, and reassured the shepherdess her sheep were alive and well.

The sheep world is a small world, and word of our successful importation traveled quickly. Within a few days, we received many inquires from people about the East Friesians including Tom Clark and Ken Kleinpeter of Old Chatham Sheepherding Company. Ken mentioned hearing a rumor that the American Sheep Industry Association (ASI) was lobbying the USDA to put a temporary ban on imports. Was this a form of protectionism? Or was the ASI concerned about the BSE situation in Europe? Unsure if or when the ban might take place, Larry and I decided to quickly arrange another importation. This would be a mixed group of sheep: mostly Beltex (from Blanchaerts and two other farms), two Charollais rams from the DeVleighers, and more East Friesians. All the animals would go to our farm with the exception of another East Friesian ewe for Mr. Freeman and two East Friesian rams for Old Chatham.

Larry once again contacted Dr. Perkins and was told to send the papers for the next importation directly to Dr. Hansen in Newburgh and just forward a copy to Dr. Perkins. The following day the second quarantine started in Belgium.

In mid-October, Larry flew to Belgium to get our next group of sheep, and they safely arrived at the Newburgh quarantine on October 20, 1996.

And finally our luck changed in our farm search. Anna Whiteside found a farm for rent, less than a mile from where we were living. The small hill farm was reached by a long winding dirt drive and had eight acres of overgrown pastures surrounded by old-growth forests of pine and spruce. The farm complex included a two-stall horse barn, a two-car garage/barn, and a large "funky" house built in the 1960s with commanding views of the valley.

Immediately we set about clearing burdocks out of the pasture, removing old fencing, and cleaning the horse barn and the garage. In the lower pasture was a beautiful pond that reflected the changing maple leaves as we worked the fields setting up fence for the sheep. The children helped out every spare moment they had, and not once did they complain about the hard work even while having to hike up and down the fairly steep slopes of the fields. To all of us, it was part of building the family business and very rewarding.

4

BENEDICT

The sheep are here! The sheep are here!" Heather yelled. I peered out the living room window through the early November darkness and saw the lights of the livestock truck at the bottom of the hill. Snow was already falling as the truck carefully maneuvered its way up the steep drive. While the driver backed up to the garage, Larry hopped out and, after a quick round of hellos and kisses, hurried to the back of the vehicle. Longing to run after their five-hour road trip, the five ram lambs leaped into the air, off the truck, and into their new home, stopping only briefly to sniff and observe their surroundings. We had converted one bay of the garage into a shelter for the rams, while the barn was thoroughly cleaned and bedded down with fresh hay for the ewe lambs.

As the truck crept back down the drive to get closer to the barn for the ewes, the children and I gathered to form a lane. "Is everyone ready?" Larry asked. "They may try to make a run for it, so be prepared." We all nodded in agreement as he cautiously opened the door.

Nothing happened.

Reluctant to leave the secure cocoon of their warm trailer, the ewe lambs stayed put. Five or six inches of snow had accumulated on the ground and little clouds floated near their mouths from the cold mountain air.

Larry climbed in the trailer and with a little prodding was able to get the first ewe lamb off the trailer. She jumped from the ramp and came to a sudden halt when her hooves hit the snow. What was it? White, cold, a little wet. This was something totally unfamiliar.

The remaining six lambs watched the whole performance and decided it was safer to stay where they were. No amount of prodding, begging, or calling would make them budge.

Impatient to try some of the handling techniques he had learned at the sheep conference in Wisconsin, Francis climbed on the trailer and attempted to move one of the sheep. The ewe lamb may not have been full grown, but she knew how to dig her hooves in, lean back, and make herself a dead weight. Francis, however, could be just as stubborn, and he finally managed to get her off the trailer. Seeing two of their companions now outside, the others quickly followed in true sheep fashion.

Heather and Jackie ran to open the gate and herd them into the warm, gently lit barn. After a quick perusal of their new home, the ewe lambs settled into a pile, ready to sleep for the night. They had formed bonds with each other over the past three months in quarantine and now seemed quite content to be together in this quiet, cozy environment. As one of the larger lambs shifted her weight, snuggling deeper into the fragrant hay, another lamb stretched and then cuddled beside her, resting her head on her back. The larger ewe grunted contentedly, and soon all eyes were closed and their breathing slowed.

I stood there in awe and amazement, surrounded by my young family, looking at our beautiful animals. The children were reluctant to go to bed that evening. So were Larry and I. We were finally shepherds.

The children were up at dawn, running out to see the sheep before going to school. When we opened the barn doors the ewe lambs made their way outside, cautiously following each other in a line. We were unsure whether they would realize there was still green grass under the soft snow, but after about fifteen minutes a Beltex lamb (soon to be known as Upsala), pawed the ground with her hoof, uncovered some grass, began grazing, and before long the others mimicked her.

The ewe lambs were all Beltex, except for one East Friesian who towered over the rest and looked unsure of why she, the regal ewe lamb, had been placed with the stubby porkers. In the ram lamb enclosure were Beltex, East Friesian, and Charollais: a true mix of shapes and sizes. The ram lambs were also bonded but tended to have little cliques. The Charollais, who were half brothers, were never more than a few feet from each other.

Larry and I spent the day checking each and every one of the sheep and making sure all their records were in order. After all the long years of planning and waiting, we soaked up the smells, sounds, and sensations of our new farm and made every possible excuse to be with them.

Naming the sheep was more challenging than we thought. Some of the Beltex ewes were named already and we didn't want the standard "Bo Peep" names for the rest, so we researched Flemish names, checked baby books, and talked with the children. Upsala, who was Miss Independent, was always the farthest from the flock, the first one to try and escape, and the largest of the bunch. Uitgefester (Ewt-ga-fester), Satra, and Olga were calm, stout, reliable, and quiet Beltex ewe lambs. Mrs. Friendly (Cara Amarita) was the easiest to name because she was always smiling and was first to greet anyone who came to the pen. Another ewe lamb was dubbed "Granny," because she seemed like a wise old soul, even as a lamb.

The rams were a little easier to name. "Fromme," our large East Friesian ram, was slang for *fromage*; "BB" was the name for the Beltex ram and was short for "Bernard Blanchaert" in honor of our Belgian friends Dr. Bernard Carton and the Blanchaerts; and "François" and "Jacques" were the names chosen for the Charollais, for Francis and Jackie.

A few days after the sheep arrived, Freddie the llama came to the farm. Most llamas will make good guard animals, but not all. Some can be too passive to be alert to dangers to the flock, and others are too aggressive to the sheep or to humans. Fellow Vermonter Gail Birutta started a guardian llama placement program to "test" llamas before they were placed with their companions, and eighteen-month-old Freddie was a successful "graduate."

Llamas are a chosen guard animal because they eat grass, have a natural hate of canines, bond to whatever animals they are with (if there is not another llama around), and will protect their charges with their lives. When we were deciding on an animal to guard the sheep from coyotes and stray dogs, we chose a llama over a guard dog because of their diet, their life span (they can live to be more than twenty years old), and the fact they would be less aggressive toward visitors. Plus, Heather wanted one.

Freddie's real name was Lithia Highlight, but when Heather saw the registration certificate she snorted, "That's a girl's name. He needs something more masculine sounding like . . . Fred, Harry, Joe, Bob . . . yeah, 'Freddie.' That's it!"

While Freddie spent most of his young life with sheep, he was accustomed to shy, skittish sheep and was unprepared for the "sheep rush" when he entered the ewe lambs' pen. The Beltex gathered around, staring up at this strange tall creature. Freddie attempted to walk backward, but a roly-poly was in the way. Never had he seen sheep that looked or acted like this. They had wool and smelled like sheep, but the resemblance stopped there!

Gingerly Freddie tried stepping away. Eventually there was an opening and he made a dash toward the barn, but the sheep ran with him. No matter which way he went, they were there. The chase continued throughout the entire day. By nightfall, exhausted, he resigned himself to the fact that these were his new penmates. When Larry and I closed the animals in that night, Freddie had two Beltex ewe lambs cuddled beside him.

The ewes quickly adjusted to the Vermont winter. Every morning Larry or one of the children would carry a hay bale out from the barn and put it in an area with new-fallen snow. This encouraged the ewes to get outside and exercise in the fresh air. As soon as the hay was down the ewes would make their way through the snowdrifts, following their network of trails. Sometimes the drifts were so high you could only see the llama.

A ewe has a five-month gestation. During the last six weeks of their pregnancy the quality of their feed is usually improved by adding grain to their diet. The Belgians warned us not to give any grain to the Beltex. The Beltex breed is so efficient at diverting nutrients to their lambs that the ewes would have lambs that were too large for easy births. We did not feed any grain, but to be on the safe side we fed the ewes rowen (high-quality, second-cut hay) the last six weeks of their pregnancies. By March the Beltex were extremely round. A little too round.

As the due dates drew near we would all go out to the barn and check on our precious flock more often. "Are they in labor yet?" I kept asking. I had witnessed calves, puppies, kittens, chicks, and rabbits being born, but never a lamb, and I was anxious, to say the least. Larry was more experienced and comfortable. "Not yet," he would patiently reply.

Finally, on March 29, Upsala showed signs of going into labor. Her udder was red and full, her vulva swollen, and she pawed at the ground, making a nest for herself and her lamb(s). She grunted and started to push.

There was a nose. A great sign! More grunting, more pushing. We stood together, watching, waiting in anticipation. More grunting, more pushing. After observing this for almost half an hour, I was concerned. Wasn't it taking too long? The nose would make an appearance, and then slide back in for a few more minutes.

It is important to allow the ewe to birth unassisted, but, this being our first birth and seeing that Upsala was not making progress, Larry suggested I feel inside to see how the lamb was positioned, because my hands were smaller. I put on plastic gloves, applied a lubricant, and knelt down in the hay behind Upsala as Larry gently talked me through the process.

All I could feel was a very large head. Larry explained how we needed to get one or both of the legs out. Even though the night air was cold, I could feel the sweat starting to run down my back. I kept trying to find the legs but was too afraid I might hurt Upsala.

When I took a break, Upsala walked around, occasionally trying to push. But the lamb did not seem to be responding when I touched its muzzle. We decided to call our vet, Dr. Karen Anderson, and within twenty minutes she joined us in the small barn, now warm with the sheep's breath and ours. After assessing the situation, Dr. Anderson pulled on elbow-length gloves and started applying lubricant.

"Get your gloves on," she told me.

I stretched on a fresh pair of gloves and lube. "Put your hand in, gently sliding over the forehead. Do you feel all of the head?"

Carefully I slipped my hand in and around the head. Upsala started pushing as she felt the pressure.

"Yes, I can feel it. It's big."

"I want you to feel for the legs, underneath the head."

As I felt my way toward the legs, Upsala leaned back, putting her weight against my shoulder. I adjusted my balance.

"I've got it. I've got a leg!"

"Okay. I want you to pull that leg out," Dr. Anderson said.

"I can't! I'll hurt Upsala."

"No, you won't. Keep going."

Dr. Anderson knelt down beside me. "I'm going to get the other leg. Keep a hold of that one."

She slid her hand in and grabbed the other leg. Within seconds she had it out.

"Now do the same with yours," she told me.

I pulled carefully, and then harder. Both legs were now out.

"Larry, why don't you pull the lamb out?" she suggested. Larry and the children had silently waited and watched the entire time.

He knelt down on one knee and pulled the lamb out, quickly hanging it upside down. There was no heartbeat. He rubbed the chest cavity. Still no response. He swung the lamb, hoping to stimulate it, but it was too late.

Larry placed the lamb beside Upsala. Licking the lamb would stimulate more contractions and help her discharge the afterbirth. Upsala softly grunted as she cleaned the lamb.

"Is he dead, Daddy?" Heather quietly asked.

Larry nodded his head sadly.

Dr. Anderson stared at the lamb. "How big is he?"

Anxious to change the subject, Larry grabbed the scale from the other room. He laid the lamb on the holder—"15.4 pounds."

Oh my God, what had we done?! A lamb should average between seven and eight pounds. Was it possible that the rowen made them too big? I looked at the other ewes, bulging like little bowling balls, and prayed some of them had twins.

Upsala pushed against Larry's legs, staring up at her lamb. She wanted it back.

As Dr. Anderson repacked her veterinary kit, she told us, "Call me again if you need me. I'll try to keep my schedule free." Larry took the lamb and walked with Dr. Anderson to her truck. I tried to give Upsala a hug, but she wanted to follow Larry.

Minutes later, Olga, one of the quietest ewes, started pawing the ground. I noticed some mucus dangling from her. She was in labor!

I ran out of the barn and up the drive, calling for Larry. Dr. Anderson had already left.

"Olga's in labor!" I yelled.

Again, as quietly as possible the family stood together, watching Olga go through the same motions.

"There's a nose!" Francis whispered excitedly.

"Do I need to pull?" I nervously asked Larry.

"No, no. Let her try on her own."

I was anxious to be of better assistance and wanted to try my new skills, but these proved unnecessary as Olga pushed out a beautiful, big ram lamb. All on her own, the way it should be.

As soon as the lamb was out, Olga walked away.

"But she needs to lick it!" I cried.

"Give her a chance," Larry said.

A few minutes later she was still ignoring her newborn. Upsala, however, was filled with the maternal instinct and kept trying to push past us to get to the lamb. When we finally relented she hurriedly waddled over to the lamb and began licking it clean.

We stood in the warm barn, surrounded by our beautiful sheep and the lamb—the first Beltex lamb ever born in the United States. The lamb lifted its head and bleated. Benedict would be its name, meaning "blessed."

5

THE PROMISE

I wanted our children to have the same experience growing up on a farm and caring for animals that I did. It was a wonderful way to teach them about responsibility, and they naturally took to farm life. Jackie spent every spare moment she had with the sheep. She was especially close with the Beltex ewes, and on occasion we found her napping beside Mrs. Friendly, Jackie's head resting on her warm, woolly back while Mrs. Friendly laid still, contentedly chewing her cud.

Heather was small and slender but incredibly strong and could carry large buckets of water and bales of hay with little effort. She took care of Freddie, and, with milkings twice a day, Heather bonded to the East Friesians. Using photos of portable parlors for milking cattle in southern Europe as a guide, Larry and I had designed a portable sheep-milking parlor for Heather. We converted a snowmobile trailer that had flip-up sides, two ramps for entering and exiting, marine-grade plywood floors, and a sunroof installed to allow in plenty of light. Most important, we made sure the height was suitable for Heather's petite stature in order to avoid repetitive injury from milking.

Francis was well rounded in all aspects of raising sheep; he learned ear tagging and sheep management skills from Larry, attended a shearing school at UVM, and studied rotational grazing so he would know how to best manage the grass on the farm. For years Francis wanted a dog, and in the spring of 1997, unbeknownst to Francis, Larry and I decided it was time. We found an adorable collie puppy whose parents were working dogs on a Vermont dairy farm.

When we arrived home Larry hid the puppy under his shirt and called Francis in a very stern voice. Francis rushed over from the lower field and thought he was in trouble. Tears of joy rolled down his face at the sight of a tail that slid out from the bottom of Larry's shirt. Being a great soccer player and an avid fan of the sport, Francis named his dog Pélé.

Not only were the children very capable of the day-to-day management of the farm but they were competent in all the business aspects as Larry and I had made sure they were included in discussions and understood the business plan.

In December of 1996 scientists speculated there was a possible link between BSE and a new variant of CJD, the human TSE, and the USDA closed the doors on the importation of live animals from Europe in 1997. In May of 1997 our company, Ag-Innovations, gained international recognition when we were the first ever to import sheep semen into the United States from Britain. Requests for our brochure and video dramatically increased.

By July 1998 our flock would be scrapie certified and we could sell animals, but in the meantime the demand was so high we decided to import live animals from Silverstream in New Zealand. We quickly had orders for a dozen sheep, and Larry and I decided I would make the journey to New Zealand to bring them home.

In early July, I flew from Vermont to Dunedin, New Zealand, where everyone at the Silverstream office was abuzz with the news that Blue 40, a Silverstream East Friesian ram, had won the New Zealand national championship for the quality of the lambs he sired and was now valued at NZ $260,000!

Silverstream also received extensive publicity for working with us to bring sheep to the United States, and I spent my first three days giving radio and television interviews and visiting sheep farms. Finally it was time to visit the sheep. Jock Allison showed me the group of animals that had been quarantined for sixty days. I inspected each sheep before finally selecting fourteen (ten pregnant ewe lambs and four ram lambs).

On Saturday morning Jock and Hilary drove me to the farm where we met the driver of the livestock trailer. The sheep were quickly loaded: ewes on one side of the trailer and rams on the other, with a wall separating both groups. Jock gave me hay (lucerne/alfalfa), "nuts" (pelletized grain for sheep), and "Ewe Life" (an electrolyte solution I could safely give the animals if any were looking worn out). After saying our good-byes I climbed into the passenger's side of the vehicle, and we set off.

"So where you from?" the driver asked as we pulled onto the main road, heading north toward Christchurch.

"Vermont."

"Now, that's part of the United States, eh?" he asked. I nodded.

"Dr. Allison told me that," he said with a proud smile, his lips pulling back to reveal a toothy grin with teeth that had seen better days. "What's the weather like in Vermont?"

I explained it was summertime when I left and how in the winter we average ten feet of snow, and much of it falls in March.

The driver let out a whistle. "I can't even imagine," he said, shaking his head.

"We live near three ski resorts, and up on the mountain they get close to twenty feet," I added, watching the astonishment expand on his face.

"Watcha doing with the sheep?" he asked, tilting his head toward the back of the truck. He normally transported horses.

"We raise sheep and sell purebred animals to other farms interested in sheep dairying."

"I didn't know you could milk a sheep until I heard about what Dr. Allison was doing, importing those sheep and all." He paused for a moment. "Must be mighty dear, moving sheep across the world like that."

I smiled. "Yes, it's expensive, but it's worth it." I explained how the typical American ewe gave only 100 pounds of milk in a year, while the East Friesians we imported from Belgium and the Netherlands averaged 1,000 pounds, and some individuals produced more than 2,500 pounds.

"You mean you've gone to Europe for these kinds of sheep, too?" he asked, trying to fathom a life that was foreign to him.

"Yep, and we also imported a meat breed of sheep called the Beltex. They look like pigs when they're shorn."

"Are they anything like the Texel?" he asked.

I looked over in surprise, not expecting a truck driver to know about a particular breed of sheep. But sheep outnumbered people in New Zealand by almost twelve to one. New Zealand had more than forty-five million sheep in an area the size of Colorado, while the entire United States had barely six million. No wonder someone not in the sheep industry would be familiar with the Texel breed.

"They are," I answered. "They were developed from the Texels in Belgium."

"Do you do anything with wool?" he asked.

"No, there are plenty of great wool breeds already in the United States, and we are more interested in the dairy and meat animals."

We drove along in silence for a few miles. I looked out the window at the rolling green hills, covered with sheep, cattle, and deer. I was surprised how many deer farms there were and asked the driver about it.

"Yeah, lamb is really popular here, but we also probably have the best deer industry in the world, over four thousand farms with almost two million head."

My mouth dropped open. I had heard plenty about New Zealand lamb but never anything about New Zealand venison.

"Is it all for meat?" I asked.

"No, the Asians like the velvet from the antlers; they say it's an aphrodisiac." He winked and asked, "Are you married?" as he looked at my hands.

I stiffened. Here I was traveling alone for three days and nights through New Zealand with this unknown truck driver, far away from Larry, transporting pregnant sheep. "Yes, I have a wonderful husband and three incredible children," I answered quickly.

"You're lucky," he said. "You sound happy." He smiled, and I relaxed as I realized the driver was just making conversation, not trying to hit on me.

"I am."

Towns were few and far between in New Zealand, and as the day wore on I became sleepy. I had started to nod off when the driver asked, "Do you need the dunny?"

I jumped and tried to shake off the drowsiness. "The dunny?" I repeated.

The driver's face turned red. "Um . . . the toilet?"

Now it was my turn to blush. "Yes, that would be good."

A few miles later we pulled into a petrol station. "Can I check on the sheep first?" I asked. The driver nodded and walked to the back of the trailer and opened the top of the Dutch-style doors. A few of the sheep were standing but most sat comfortably in the bedding. I took out the small bucket and poured one of the bottles of water Jock had packed and offered it over the door to the sheep. None of the ewes were interested, and only one of the rams walked straight over to me, sniffed the bucket, and then noisily sucked up some water.

It felt good to stretch my legs, and I noticed my stomach was rumbling. Inside the small shop I bought a bottle of water and a packet of crisps. As I climbed back into the truck, the driver saw my purchases. "Would you like something to eat? I knew you would be traveling with me and packed some extra sammies," he said as he pulled out a few simple cheese sandwiches from his "chilly bin." I accepted one gratefully.

The light started to fade a few hours later, and I figured we were getting close to the northern tip of South Island. The driver turned off the main road and maneuvered the trailer through a small residential area and then into a more rural setting.

"We'll be stopping here," he said as we pulled alongside a barn and a few houses. I looked around. The last hotel I had seen was shortly after lunch. "Must be a B&B," I thought.

The driver got out first and ran around to open my door and offer an arm to help me down. "Will the sheep be staying here all night?" I asked nervously. New Zealand appeared to be a safe country, but the thought of leaving our valuable purebred sheep alone, parked by a barn, made me nervous.

"This is an authorized holding area," the driver answered. "I do this all the time, but mostly with horses."

I looked around and could only see a few houses and open fields.

"Can I water and feed the sheep first?" I asked.

"Sure," he said, walking around the back to again open the doors, this time all the way so I could get in the pens with the sheep. Most of the sheep were calm, but a few stepped away from me as I entered. Jock recommended feeding a kilo of "nuts" twice a day and gave me a scoop that would hold the suggested amount. One by one I fed them and made sure they each had their fill of water. Then I gave one and a half flakes of alfalfa to the rams and three flakes to the ewes. This would give them plenty to eat through the night. When I was satisfied that all the animals were taken care of, I stepped out and the driver closed the doors. "Do you have any more bags?" he asked, grabbing my suitcase from the cab.

"No, just that one."

"You sure do pack light. I thought women liked to travel choc-a-block."

I laughed. "No, I wanted to be able to take care of the sheep and not have to worry about lugging around a bunch of luggage."

As we approached the front door of the house I noticed there were no lights on. The driver took out a set of keys from his pocket and opened the door, holding it for me to enter. As we walked in he tossed the keys on a small table near the door and ran ahead to turn on some lights.

"Would you like something to drink?" he asked.

I stood in shock in the entranceway. This was not a B&B; this was his house, his bachelor pad—a pretty tidy bachelor pad—but a single man's home nonetheless. What was I going to do now?!

"Um . . . no thanks. Do you have a hotel or B&B I could stay at?"

The driver looked genuinely surprised. "But you're supposed to stay here."

"What?!" I exclaimed.

"There's not a place to stay around here unless you go all the way into Picton."

"Oh," I said quietly.

"Didn't Dr. Allison tell you? Grooms riding with the horses generally stay here," he said motioning toward a large couch with blankets and pillows stacked neatly beside it, "unless they have some mates nearby, then they stay with them."

"No, he didn't mention it."

The driver stared at me for a few moments, noticing my blushing cheeks and overall nervousness. "Look, if it will make you feel more comfortable, my mum lives next door. I can stay there. You are welcome to sleep in my room."

"No, no, that's fine, I'll sleep on the couch," I said quickly, not wanting to displace him from his own home.

The driver looked disappointed. First I didn't want to stay in his house, and now I was refusing the offer of his room.

"Would you like a cuppa?" he asked.

"Sure, that would be wonderful."

The driver perked up. "Milk and sugar?"

"Yes, please."

"Have a seat in the living room, and I'll bring it out to you."

I sat nervously on the edge of the couch. This was not what I'd expected when I planned this trip. The house was very quiet with only the sound of a clock ticking on the wall. I looked at the magazines scattered about, mostly sports and a few car magazines. I found one on rugby and started reading. Larry played rugby in college and taught me to enjoy the game.

The driver appeared and set down a round metal tray with two piping hot cups of tea, some sugar in a small Vegemite jar, milk in a cracked creamer, and a plate of "biscuits" on the coffee table. Immediately he began apologizing, "Sorry I only have some digestives. I've been traveling a lot, and there's not much in my kitchen. Is sugar okay? I could try to rattle up some honey if you want."

"No, no, this is fine. I really appreciate it," I said, pouring some milk into the warm dark liquid.

"Well, it's not every day I get to entertain someone from America," he said smiling, handing me a teaspoon for the sugar.

We ate the biscuits, drank the tea, and talked about the All Blacks' recent win "across the ditch," as the driver said. I told him about Larry coaching the Virginia Tech women's rugby team and how they won the state championship. "And he played in England and coached the University of Nottingham's women's team for a short while," I said. "Even our son, Francis, played rugby. In his first game when he knocked down a player from the other team he stopped and helped him up, apologizing the entire time. His coach yelled at him to keep going."

The driver laughed, "Well, how did he do?"

"He scored a try!"

"And Bob's your uncle! Good for him!"

The tea was gone, most of the biscuits were eaten, and I started yawning. The driver, ever attentive, hopped up from his chair and began clearing the dishes. "Can I help you wash these?" I offered.

"Oh no, it's fine. I'm just going to set these in the kitchen and will do them in the morning. What time would you like me to wake you?" he asked. "The driver will be ready to leave by 7:00," he said in anticipation of my question.

"You won't be driving?"

"No, no, the driver tomorrow will take you on the ferry to Palmerston North."

"Oh, could you wake me up at 6:30 then?"

"Sure. Now, I'll go get my room set up for you."

"I can't take your room," I insisted.

"It's okay. You need a good night's sleep with all the days of traveling you have ahead of you. And I have tomorrow off. No driving for me on Sunday!" he said happily.

"I appreciate it, but the couch is fine. I'll just sleep here."

He actually looked relieved. Maybe the thought of cleaning his room late in the evening was not his idea of fun. "All right, if you insist. I'll be back in the morning to wake you up. G'night."

"Yoo-hoo, Miss, it's time to get up!"

I rolled over and opened my eyes to the smiling driver standing over me.

"Cracker of a day, eh?" he said, opening the curtains to the morning sun. It did look to be a gorgeous day. When I started to get up from the couch, the driver rushed toward the kitchen. "I'll put the kettle on while you get ready."

There was no need for him to be embarrassed—I slept in my clothes. After a quick cup of tea we walked to the barn. As we turned the corner, I gasped. The livestock trailer was gone!

"The sheep!" I cried, "Where are the sheep?!"

The driver laughed and pointed to a larger livestock trailer now parked near the front of the barn. "They're in there."

The sheep were already loaded, again in two separate pens, but this time they shared their accommodations with six horses in adjacent stalls. "Let me introduce you," the driver said, walking toward a man and two women. Introductions were made with the usual handshakes, and soon it was time to get on the road. I turned to the smiling driver and extended my hand. "Thank you for all your help," I said.

He grasped my hand and shook it enthusiastically. "No problem. Glad I could help. Now you take good care of those sheep, you hear me?"

I laughed, "I promise. These sheep are going to wonderful farms where they will be very well taken care of." I could see them living out their lives with their lambs growing beside them in the green fields of New England. They would—I promised.

6

LOSS OF INNOCENCE

July 14, 1998, was a beautiful, warm, sunny day, and I was ready to relax. After five days of traveling from Dunedin, New Zealand, to Newburgh, New York, with our fourteen sheep and then home to Vermont early that morning, I could think of nothing better than to enjoy the summer day with my family. As I stood on the deck and saw our sheep serenely grazing in the fields, I felt tremendous pride. The sheep looked great, the farm looked great. Iced tea was in the fridge and a plate of farmstead cheese was ready for serving. A gentle breeze was blowing, and the sky was a brilliant blue. Surely Dr. Detwiler was coming with good news.

A few weeks before leaving for New Zealand, I had received a phone call from Wayne Zeilenga. "Dr. Linda Detwiler would like to meet with you in July."

"No problem," I said, assuming it was about the scrapie program. Our farm was the third flock registered in the Vermont scrapie program, Mr. Freeman's was the first, and another farm that imported sheep from Canada was the second. Because the European parents of our sheep had been enrolled in the EU scrapie surveillance program since 1992, we would receive our "certified" scrapie status after being in the program for two years instead of five. This would allow us to sell sheep as soon as we received the papers. And we were due to receive our scrapie certification this month.

Dr. Detwiler was a strong supporter of the scrapie surveillance program and touted it in her article "History and Control of Scrapie":

> The program provides participating owners with the opportunity not only to protect their sheep from scrapie but to enhance the marketability of their animals. The control effort focuses on risk reduction and sound husbandry practices. Since each advancing phase represents a lower risk of scrapie

in the flock, the economic value of the animals is increased, especially after completing the five-year program and attaining "certified" status. This program may also have implications for exporting breeding stock to other countries.

Larry was now vice president of the Vermont Sheep Breeders Association and gave a talk about scrapie (including updates on current research and the national scrapie program) at the annual meeting. He explained that enrolled flocks would have each animal tagged with official tags, and once a year the USDA veterinarian for Vermont and the state veterinarian would visit the farm, inspect the flock, confirm that all animals were tagged, and if any animal exhibited any neurological symptoms have it necropsied. The scrapie program was a great way to reduce and possibly eliminate the disease, and Larry encouraged all VSBA members to join.

Wayne was always smiling and quick with a handshake, while Sam was more reserved when they visited our farm each year for our scrapie surveillance program "checkup." The process was always the same: after Wayne and Sam donned their gear and sanitized their rubber boots, we headed to the barn where Francis had the sheep in different pens and was able to quickly capture each and hold them still while Sam read their tags and Wayne marked the numbers on his sheet attached to his clipboard. The entire inspection lasted less than an hour.

Dr. Linda Detwiler was a senior staff veterinarian at USDA/APHIS and was very involved with the sheep industry. Now she was coming to see our operation, I told Larry, to congratulate us on doing such a great job: importing excellent, healthy sheep from Europe and New Zealand, getting the scrapie program up and running in Vermont, and stimulating the sheep industry, particularly the dairy sheep sector. But Larry disagreed. Something was niggling at him. "This does not feel good," he said.

They arrived on time—Wayne, Sam, Dr. Detwiler, and Dr. Bill Smith, the USDA New England regional veterinarian—and we greeted them on the front porch accompanied by Dr. Karen Anderson, whom Larry had the foresight to invite.

Wayne was grinning as usual, very proud to be with people he felt were of importance. Sam was quiet, wanting to stay in the background.

A little quieter than usual, I noted. Dr. Smith cautiously smiled, offered his hand, and introduced himself. He had dark hair and dark eyes and was not much taller than me.

Dr. Detwiler was younger than I expected. In her midforties with short, light-brown hair and a slender build, she extended her hand for the introductions, but her handshake was very short. When I looked in her eyes they were blinking rapidly, and she was not smiling. My heart sank, and a feeling of foreboding made my palms sweat. Trying to shake off the negative feelings, I smiled and invited everyone to sit on the back deck.

As soon as everyone was settled, Dr. Detwiler got right to the point. "We are concerned that your sheep could have come in contact with contaminated feed and possibly be susceptible to BSE. We have already spoken with Mr. Freeman, and we would like you to surrender your animals."

"What?!" I exclaimed, completely taken off guard. This was not the congratulations I anticipated. Surely this could not be real.

"Wait a minute!" Larry said, "There's nothing wrong with our sheep. We have the feed records to show they were never fed any feed with meat and bonemeal, let alone contaminated feed." Wayne was no longer smiling, and neither was I.

"It doesn't matter," Dr. Detwiler replied. "There is new research which shows that sheep can contract BSE."

Larry and I looked at each other: "new" information? The experiment demonstrating that sheep and many other mammals could contract BSE under laboratory conditions was published in the *Veterinary Record* on October 10, 1993—*five* years ago. Three years *before* we even imported the sheep.

"Are you referring to the *Vet Record* article?" I asked.

"Yes."

"That was published in October 1993."

Flustered, Dr. Detwiler said, "No, no, this was more recent." She sat up even straighter in her chair. "These sheep could be at risk. We can't take that risk."

Who was the royal "we" she was referring to, I wondered. Who was behind this besides the USDA?

I was jolted back to the conversation when Larry said in an annoyed tone, "No sheep in the world has ever had BSE."

"Yes, but I have information I can't divulge which would change your mind," Detwiler said firmly. The meeting was not going the way she intended. She had no idea what our backgrounds were and probably wondered why we would dare to question her.

"We will not surrender our animals," Larry said resolutely.

"Our animals were lambs when they were imported," I added, "and they were born *after* the European ban on meat and bonemeal and two full years *prior* to the U.S. ban."

Detwiler said nothing and looked at Wayne and Sam. Wayne smiled hesitantly back, while Sam quickly looked at the ground. Detwiler looked at Bill, but he had nothing to say either. The seconds dragged as the weight of the situation sunk in.

Finally, with the sound of resignation in her voice, Detwiler said, "We need time. Would you agree not to sell any sheep until we have this information?"

"How long will that be?" Larry asked.

"Well, I'll be attending another meeting in Europe with scientists who study these things," she said proudly.

"You mean the SEAC meeting?" I asked, unable to keep the edge out of my voice. Her attitude was getting to me. We were not a bunch of "country bumpkins." I knew she was an invited guest on a subgroup of the SEAC (Spongiform Encephalopathy Advisory Committee), an organization that studied TSEs. Detwiler religiously attended all the meetings.

"Yes, the SEAC," she said. A look of surprise flashed quickly across her face, then just as quickly the impenetrable look returned. "So will you agree to not sell any sheep before then?"

In a few months it would be breeding season—prime time for selling rams. We had some very nice rams available, and customers from around the world wanted to purchase them for breeding stock—Americans, Canadians, and New Zealanders were interested in the Beltex, a group of Mexicans requested a quote for a large number of East Friesians, and Jock Allison talked about a joint venture between Silverstream and Ag-Innovations. Now was the time to be selling, not to be putting everything on hold. But what choice did we have?

Reluctantly, Larry and I agreed.

• • •

During the meeting the girls had played in the house while Francis worked outside with the rams. Once the meeting ended, the girls brought out a plate of cheese and kindly offered it to our visitors. Dr. Smith smiled and took some. Dr. Detwiler stopped, looked at the cheeses and then back at the girls, and refused. Wayne and Sam followed her lead.

Francis walked inside with Pélé close on his heels. "Would you like to see the sheep?" he asked enthusiastically, anxious to show them off and unaware of what had just happened.

"Sure," said Dr. Smith. And they walked to the ram's paddock.

"Wait! No!" Detwiler cried out. "We have to be going."

But it was too late. Dr. Smith was already by the fence looking at the rams from a distance, as Francis pointed out the three breeds and talked about their personality differences. "That's Fromme, the tall East Friesian ram. He's the boss, but BB, the Beltex ram, doesn't let anyone push him around. He's not aggressive, just confident. François and Jacques are brothers and always hang out together. They're the Charollais rams."

Detwiler waited impatiently to leave. She wanted to be away from this place—the sooner the better. She never laid eyes on a single one of our sheep.

As soon as the vehicles were no longer in sight, we called Mr. Freeman. He had met with the group the previous day and was told the same thing. Linda Detwiler asked all of us if the USDA could have any animals we were going to cull. We all agreed, wanting to prove our animals were free of TSEs. Mr. Freeman told her he had a ewe with severe mastitis. (Mastitis is a common affliction of lactating mammals that results from a bacterial infection of the mammary gland and can render the animal completely unproductive.) The Whittens had tried various treatments to heal her but were unsuccessful. Mr. Freeman gave the ewe to the USDA.

Chet Parsons was in charge of the UVM booth at the Champlain Valley Fair and invited Larry and me to do a cheesemaking demonstration on September 3, 1998. He thought the presentation would be a good way to teach people about cheesemaking and milking sheep. A member of the Vermont Scrapie Certification Board and the VSBA board, Chet was also a fellow sheep farmer and our extension agent. With gray hair and beard, a great singing voice, and a wonderful smile, Chet was adored by all in the sheep world. It was another beautiful day when Larry and I arrived at the booth and set up the display, and Chet was grateful to have the company. There was a large turnout, and the booth was well received.

Around 4:00 in the afternoon a roundish man with orange hair and a boyish face rushed over to the booth. "Did you hear about Mr. Freeman's sheep?" he asked Chet breathlessly.

Immediately Larry and I stopped and listened. "What about Mr. Freeman's sheep?" Chet queried.

"They have scrapie and might have BSE!"

"Who is this guy?" I whispered to Larry.

"That's Leon Graves, our Vermont commissioner of agriculture."

"What is he talking about?" I asked.

Larry quickly hushed me. "Shh, so we can hear what he's saying."

Leon glanced over at Larry and me, quickly assessed us, and decided we were not important. "There was a test done on one of his sheep and it came back positive for scrapie and maybe BSE!" Leon continued animatedly telling Chet.

Chet stood there quietly, wondering if Leon knew who we were and, judging from our body language, understood we wanted to remain anonymous.

Leon made small talk for a few more minutes and then scurried off, as excited as when he arrived.

"What was that all about?" we asked Chet.

"I have no idea."

"What the hell is this about Mr. Freeman's sheep testing positive for scrapie?" Larry yelled over the phone at Wayne.

"What?!"

"Leon Graves was at the Champlain Valley Fair today telling Chet Parsons and anyone else who could hear that one of Mr. Freeman's sheep tested positive for scrapie and maybe BSE."

"Oh no! Calm down, calm down, Larry. At first we thought there were some signs. But numerous tests were run, and they were all negative."

"What do you mean 'signs'?"

"From the histopathology. But all the tests were negative."

"So why is Leon saying this?"

"He must not understand. We will straighten this out."

Scientific understanding to date is that TSEs start in the intestinal system, make their way into the lymphoreticular system, and, after a *minimum* of two years (twenty-four months), up the spinal cord and into the brain stem. When testing for scrapie, BSE, CWD, or any of the other TSE diseases, a histopathology is the first procedure performed. For the histopathology, scientists slice off a piece of the brain stem (the medulla), place it under a microscope, and look for three distinct cellular changes:

1. Astrocytosis—Astrocytes are star-shaped cells that support and nourish neurons (brain cells). Proliferation of astrocytes (astrocytosis) occurs when nearby neurons are damaged. Astrocytosis appears whenever an animal has a fever, inflammation, or is stressed. Therefore astrocytosis is seen on many slides.
2. Necrotic neurons—Neurons are constantly dying and being regenerated, but in a sick animal the neurons are dying and not being replaced; therefore, numerous necrotic neurons are observed in a sick animal, while only a few are seen in a healthy animal.
3. Spongiosis—This is where the term *spongiform* originated. Applying dye to the slide with the slice of brain stem can reveal holes that make the brain appear "sponge-like." This is a defining characteristic of TSEs.

For an animal to be considered "suspect" of having a TSE, all three changes must be observed. According to an article written by Drs. Richard Race, Darwin Ernst, and Diane Sutton, "Individually, none of the expected microscopic lesions, which include neuronal degeneration, neuronal vacuolation, astrocytosis, and spongiform degeneration, is pathogenic for TSE." (And Diane Sutton worked for the USDA in Ames, Iowa, at the laboratory that tested this ewe.)

If all three changes are identified, an animal is considered "suspect," and additional confirmatory tests are run. If all the changes are not observed, the animal is declared negative for any TSEs. As would be expected, Mr. Freeman's ewe had a minor amount of astrocytosis because she had severe mastitis (an inflammation of the udder) and a fever. There were no necrotic neurons and no spongiosis; therefore, no TSE.

So was Leon mistaken or was he spreading rumors?

7

CAUGHT IN THE MAELSTROM

The USDA requested a list of all the people we and Mr. Freeman had sold sheep to and then proceeded to contact the farms and issue an ultimatum: either surrender the European animals they purchased from us, or their entire farm would be quarantined. Needless to say, our customers were upset, concerned, and angry; and many thought we had done something wrong. Why should they be forced to sell their animals? Didn't we have full USDA approval to import the animals from Europe?

We called Ken Kleinpeter of Old Chatham Sheepherding Company in New York and suggested he have Roger Ives collect semen from their rams. A few days later Ken told us, "USDA didn't want to purchase the rams at first, but now they do. They said I could buy a few weeks for semen collection, but I don't want to have my rams killed if you guys are able to save your sheep with a deal." We reassured him we had no intention of selling our irreplaceable sheep, and we would keep him posted.

In 1997 the European Parliament established the Scientific Steering Committee (SSC) to assess each country's BSE risk and risk management. The SSC developed a questionnaire and the answers were then analyzed, and each country interested in exporting meat was ranked on a scale of I to IV for their risk of finding BSE in their country, with I being "Highly unlikely"; II "Unlikely but not excluded"; III "Likely but not confirmed or confirmed, at a lower level"; and IV was "Confirmed, at a higher level." England was a level IV.

The methodology for calculating a country's ranking was based on eight factors:

1. Structure and dynamics of the cattle, sheep, and goat populations
2. Animal trade
3. Animal feed and import of animal feed

4. Meat and bonemeal (MBM) bans
5. Specified risk material (SRM) bans
6. Surveillance of TSE, with particular reference to BSE and scrapie
7. Rendering and feed processing
8. BSE- or scrapie-related culling

Europe had banned the feeding of meat and bonemeal nine years earlier, did extensive testing for BSE, and kept certain parts of the animals (referred to as "specified risk material," or SRM) out of the human food chain. The United States lagged woefully behind.

There were no bans on feeding MBM or SRM, and as far as imported MBM, U.S. officials claimed that "no MBM imports from Europe occurred between 1990 and 1998." However, after further investigation the Scientific Steering Committee discovered that the United States imported thirty-six tons from the UK in 1997.

And the United States not only had scrapie, CJD, and a TSE found in farm-raised mink (transmissible mink encephalopathy, or TME) but was the only country to have CWD. The U.S. was given a level II for BSE risk.

The BSE risk assessment was discussed at the SEAC meeting on September 30, 1998. Nothing in the risk assessment for any country referred to dangers from sheep, and in their written public statement SEAC reiterated that no sheep had ever naturally contracted BSE but that the committee members were concerned that sheep which were given meat and bonemeal were at risk. They had discussed this same concern over the past ten years. Because meat and bonemeal was added to some commercial sheep feeds in the United States, all American sheep would be considered at risk. The minutes from the SEAC meeting reviewed the possibility of BSE in the EU sheep population and concluded that more research was needed and "the Committee agreed that, at present, there was no need to recommend further action to protect public or animal health."

Larry and I breathed a sigh of relief. We had the feed records and knew that none of our imported animals or their relatives ever received meat and bonemeal. We called Mr. Freeman and reassured him that the situation would soon be over. Next we called Wayne to arrange a meeting to resolve the entire matter. He had still not heard from Linda Detwiler and said he worried about "the wishy-washy department dragging this on."

"What a strange response," I thought.

Montpelier is the capital of Vermont and the smallest capital city in the United States. With a population of only nine thousand, you would be hard-pressed to even call it a city. It is a quintessential small New England town with its white, wooden, steeple-top churches and tree-lined streets and sidewalks. It's a place where people know their neighbors and stop to talk to strangers on the street. Two main streets—Main Street and State Street—form a T in the town. Lining Main Street are small family-owned businesses: bookstores, hardware stores, clothing stores, gift shops, a sports store, antiques shops, an office supply store, offices, a few banks, a few churches, an amazing library, an independent movie theater with one small room for viewing films, and, because Vermonters appreciate good food, a variety of excellent restaurants—including two that showcase the talents of the resident New England Culinary Institute.

State Street has a few shops and a hotel, but most of the street is occupied with tastefully built and landscaped government buildings, the focal point being our state capitol with its expansive manicured lawns and gold glittering dome. Across the road from the capitol in a large, gray, stone building is the Department of Motor Vehicles; and beside it is the Department of Agriculture in a smaller, but still impressive, red sandstone building with turrets on either side. This is where Detwiler finally agreed to meet with us on October 8, 1998.

Our meeting was held on the third floor, and the atmosphere was surprisingly tense as we entered the room. Sitting around one end of the long conference table were Wayne, Detwiler, Leon Graves, Bill Smith, Dr. Russell Laslocky from the Vermont Department of Agriculture, and the Vermont state attorney. Larry and I invited my father, Dr. Glenn Cahilly, to attend the meeting with us. We sat at the opposite end of the table with Mr. Freeman, his lawyer (Tom Amidon), and Frankie and Marybeth Whitten.

Detwiler opened by talking about the EU meetings. "I've been doing a lot of traveling. I was in the Azores a few weeks ago and then in the UK. I have it down where I can be in London in the morning and back home that night," she boasted. I bristled. We didn't want to hear her travel schedule, we wanted to finish with this meeting and get on with our lives.

"As you know, I am on the European SEAC committee. There are two concurrent things going on with the subgroups: The Steering Committee is basing their decisions and actions on public perception and politics, and the Working Group is basing theirs on science."

"Was there anything new regarding BSE in sheep?" Mr. Freeman asked, already knowing the answer.

"Well, scientists are concerned that sheep that have eaten meat and bonemeal could have BSE, and it might appear as scrapie," Detwiler replied.

"We already knew that. They have been looking for years," Larry said, a heavy sigh in his voice.

"But scientists are concerned—" Detwiler said.

"But there is no new information to report," Mr. Freeman interjected.

"Your situation has been brought up a whole department level," Detwiler replied, trying to ignore the interruption.

"But there is no new information then," Mr. Freeman repeated, this time a little louder.

Detwiler paused.

"The SEAC minutes said that no further action would be taken on the sheep population," I inserted.

"Yes, that is what the minutes said, but I have information I can't divulge," she replied.

"What information?" Mr. Freeman pressed her.

"I'm sorry, I'm not at liberty to discuss it," Detwiler responded.

"We will offer full market value for all sheep and progeny," Detwiler continued. "The president of the American Sheep Industry Association and the director of research and education came to my office. They said if the USDA didn't kill the sheep, they would go before Congress themselves to get the appropriate funds to buy the sheep."

"What?!" I exclaimed.

"We are not going to sell our sheep," Mr. Freeman said angrily. Larry and I vigorously nodded our heads in agreement.

"There is nothing wrong with our sheep, and no new information out of Europe, so why is this continuing?" Larry demanded.

"I know more than I can say," Detwiler said.

"Realize that there are forces above us that we have no control over,"

Wayne added. All eyes were now on him. Detwiler sent him a repri-
manding look.

"What forces?" I asked.

"I can't say," Wayne said, trying to back away from the issue.

"Forces within the USDA or from outside the USDA?" I asked.

The room was silent. Finally, Wayne spoke, "Just remember, perception is reality."

What was this all about? Why were they speaking in code?

"What about sheep imported through Canada; what about CWD
and deer and elk farms? Why aren't you going after them?" I asked.

"As far as the sheep from Canada, if you give us the names of people
you know who imported sheep, we will review the situation," Detwiler
replied, trying to maintain her composure.

"Yeah, right," I thought, "as if we would help you on your witch
hunt." We knew people who were milking Canadian East Friesians,
but the last thing I wanted was for anyone else to be subjected to this
insane ordeal. The USDA seemed to be relentless with us and anything but thorough with the overall picture.

"So you're not doing anything about those sheep, but you want us to
surrender ours? This is ridiculous!" Mr. Freeman exclaimed. "What
about a compromise? Surely there must be a compromise, something
we could all agree upon."

Detwiler refused to budge. "We are under political pressure."

"From whom?" I asked again.

"We can't say."

"What can you say?" Larry asked with a sharp edge to his voice.

Detwiler sent him a glaring look. Ignoring his question, she continued, "It's important to not tell anyone about this, especially the
media. If this got out we wouldn't be able to control it. Will you agree
to keep the voluntary quarantine for now?"

An uncomfortable silence filled the room. We looked at Mr. Freeman,
he looked at Tom Amidon. Reluctantly, we all agreed—again.

Another meeting of getting nowhere. Why was the USDA doing
this? Surely they couldn't force us to surrender perfectly healthy animals. Larry and I stayed current on all the scientific information from
Europe, and yet the USDA was trudging forward—ignoring science,

playing politics. Mr. Freeman was noticeably silent as we walked down the stairs. We were not used to seeing him look discouraged. Here was the man who believed in us, and we didn't want to let him down or to lose everything we all worked so hard for. No one spoke until we were on the sidewalk.

"You need to gather as much information as possible from Europe," Mr. Freeman told Larry and me. "I'll fly over anyone who might be helpful. I'll be at the New York office for a few days, but let's stay in touch."

The ride home was unusually quiet. After we left my father at his office, we drove past the sheep. The ewes were grazing, and the lambs were sleeping on top of each other. What would this bucolic scene be like next week, next month, next year?

The next day when Larry and I met with Linda Detwiler, Bill Smith, Leon Graves, and Bob Paquin at Senator Leahy's office, Detwiler had a slightly different tone. "Can you get us the source flock records?"

"We can get you any information you need," Larry replied.

"We would welcome that," said Detwiler. "We are going to formalize our offer next week and will put it in a 'decision memo.' The USDA recognizes yourselves and Mr. Freeman as being cooperative and acknowledges that there have been no violations on your part. We need to use the utmost care with this situation."

"But I thought that you found scrapie!" Leon Graves piped in.

Detwiler glared at him, and like a mother correcting a child for a repeated mistake, she said, "This action is not based on that animal."

Leon sat back in his seat and remained quiet the rest of the meeting, sufficiently reprimanded.

Bob Paquin saw the power Detwiler was wielding over Leon and said, "The senator wants his constituents treated fairly. No squeeze play. Treat them in a professional manner."

Detwiler nodded.

Bob continued, "Vice President Al Gore has contacted the Vermont Department of Ag and wants the official position of the USDA."

"We will take care of the White House," Detwiler said, sitting up even straighter. She was quiet for a moment and then said, "We want to buy the animals and will pay fair market value."

"What *is* 'fair market value'?" Larry asked. "You keep saying 'fair market value,' but we have no idea whether you are talking about $100 an animal or $100,000."

Detwiler and Bill Smith discussed this between themselves for a few minutes and finally Detwiler said, "I'm not sure how to attack this issue. Maybe we need to do it legislatively. But the one thing we need to agree on is someone to do an appraisal."

"What about collecting embryos?" Larry asked.

"I don't know."

"What about quarantining the animals and buying them as they reach old age?"

"I don't know."

"What about importing Canadian embryos?"

"I don't know. We will address these concerns in the memo."

Detwiler then repeated what she said at the meeting the previous day. "It is very important to keep this out of the media. If you are questioned by reporters, bill it as agencies and farmers working together to address a remote possibility . . . out of undue concern . . . the quarantine is to make certain."

She continued, "Bill the actions as the ultimate effort on behalf of this developing industry. You need to be fully rational in a thirty-second sound bite. We will help you through this."

"Exactly what kind of help are they thinking of?" I wondered.

Bob Paquin spoke up, looking directly at Detwiler, "Again, the senator wants everyone treated fairly."

"The USDA won't do anything arbitrary," Detwiler assured him with a smile.

The windshield wipers were mesmerizing as the rain fell on the truck. Another day, another meeting, more feelings of gloom and doom. "I think it's time to tell the children," Larry said as we drove home. I felt a sense of finality as he said those words.

An hour later the children were gathered around the dining room table. Half-packed boxes were strewn everywhere. In the midst of everything, the woman who owned the farm we were renting had stopped by to check things out, saw how nice the farm looked after we had managed the pastures with the sheep, and had decided she wanted to move back with her horses. Luckily, again with the aid of

Anna Whiteside, we found a small house to rent a little over a mile from where the sheep were now pastured.

Because most of the meetings and phone calls with the USDA took place while the children were at school, they were unaware of how serious the situation was. Larry and I had been convinced the whole thing would "blow over" and life would get back to normal, but now we weren't so sure.

As we sat at the table I tried not to cry, looking at their young faces, expectantly waiting. Francis was now fifteen years old, and Heather and Jackie were thirteen and twelve, respectively.

"The USDA says we have to sell them our sheep," Larry started.

I watched his handsome face, twisted with sorrow. This was the man I would spend the rest of my life with—who shared my hopes and wishes, the most intelligent man I know. The man who made me laugh, made me cry, and encouraged me to dream. He was a wonderful father, always putting his family first—pushing the children to do their best and encouraging responsibility. As he sat there talking to the children as equals, I fell in love with him even more.

"What do you mean sell the sheep?" Francis asked.

"The USDA said they will pay 'fair market value' . . ."

"What's their definition of 'fair market value'?" Heather demanded.

"We don't know," I replied.

"There's nothing wrong with our sheep!" Jackie cried out.

Stunned, everyone stopped and looked at her. Jackie was known as the quiet one in this family of noisy Irish Italians. "We won't sell them," she continued staunchly, folding her arms on her chest for emphasis.

"I'm sorry, Honey, but the USDA is forcing us," I told her.

"How can they force us? They don't have the right," Francis said angrily.

Heather started to cry.

"Realize that the USDA could pay us what they are worth and we could start over, maybe with more sheep or with another project," I said halfheartedly.

The children looked at Larry and me as if we had just suggested killing their best friends for money. They were unanimous: It didn't matter if the USDA offered us millions; it was not worth selling our sheep. Everyone sat in silence, not wanting any of this to be true.

Finally, Francis spoke up. "We worked hard to get where we are."

"It's not fair to the sheep," Heather added.

Jackie sat there quietly, arms still folded. She was planning her strategy. She would straighten out this whole mess.

The next morning, the following handwritten letter was on the dining room table:

> Urgent!
>
> Dear Mr. President,
>
> My name is Jackie Faillace. I am in need of your help. I am a twelve year old girl, and I live in Warren, Vermont. I own 55 sheep that were imported from Belgium and the Netherlands in 1996. My sheep are very important to me and my family. Every morning my whole family gets up early and does all the sheep chores.
>
> The USDA is trying to take away my sheep. They are trying to take my sheep away because they say there is a possibility my sheep have BSE. I know that my sheep don't have BSE. First of all there has never been any sheep that have naturally gotten BSE. The only exception was in a lab test when BSE was injected into a sheep's brain and when they fed BSE infected brains to a sheep. Still then it was extremely hard for the sheep to catch BSE. Our sheep don't go around eating BSE infected brains or get BSE injected into their brains.
>
> This is not Fair [sic] at all. The only reason USDA is doing this is because a lady named Linda Detwiler said, "We have to get rid of those European sheep even though we know there is nothing wrong with them." I thought the United States was a country where everything is equal and fair. If the United States is still trying to be a fair country you will help, and stop the USDA from taking my sheep.
>
> Sincerely,
>
> Jackie Faillace

I put the letter in an envelope and mailed it.

• • •

I was on the phone so much my ears hurt. I called Professor Lamming, assuming that if there was a problem with BSE in sheep it would show up in the UK first. Because everyone in Britain was still tender from the "mishandling" of the BSE crisis, surely scientists and politicians there would err on the side of caution. Prof had nothing new to report. Neither did friends of ours at the Scottish Agricultural College in Scotland. When I called another sheep expert from Britain, Dr. Ian McDougall, he was out of the office, but the man who answered the phone said there was nothing new in the sheep world. "People are more interested in what your president is doing than about sheep," he said, referring to the Clinton/Lewinsky scandal.

Our Belgian friends felt the disappointment and frustration of our situation. "Try to collect embryos from all the lines," Freddie Michiels, the Belgian cheesemaker who taught us to make cheese, urged us.

When I asked Mr. Pissierssens, a Belgian dairy shepherd with more than forty years of international business experience, if he thought Belgium might be able to help out, he cautioned, "Realize there are agreements between countries that prevent them from interfering with each others' policies."

Carton said all was quiet in the Belgian sheep world; the country was dealing with flooding problems and most shepherds were busy evacuating their sheep from the lowlands. He also mentioned that he had received the test results from Mr. Freeman's mastitic ewe and sent them to a pathologist friend who reviewed the results and said he would put his reactions on paper. The pathologist told him he couldn't believe the USDA tried to create hysteria out of nonspecific results.

The next day I was able to speak with Dr. Ian McDougall. Again, nothing new. "SEAC seems more interested in investigating sheep for BSE than cows. You know, there was no increase in scrapie cases when there was an increase in BSE cases," he said. This helped reassure some of the scientists who were concerned that BSE could be masking itself as scrapie in sheep.

"The lamb and pig industries are really suffering right now," Ian continued. "Last year farmers were getting £30 to £40 per ewe, this year £3."

Back in Vermont, Bob Paquin was a bit more upbeat. "The USDA seems amenable to allowing you to collect embryos from your sheep," he said. "I've been talking with Belgium and there are many sensitive EU/U.S. trade issues, and some are semirelated." He couldn't be more specific but promised to keep us informed.

When I spoke with Prof again he told me the British newspaper headlines were "English Sheep Don't Get BSE."

"I think the chance of your sheep having BSE is about the same as a meteor falling on your front porch tomorrow!" Prof said. He suggested we contact Al Gore's office directly. "This has to be based on international trade disagreements," he said. "Maybe it's the old cattle/sheep wars."

As I hung up I realized the story was getting more complex. Was this retaliation for the EU not accepting U.S. hormone-treated beef? Was someone in the sheep industry angry over our importation? Was the dairy industry feeling threatened? And what was the political pressure Detwiler referred to? The National Cattleman's Beef Association? The pharmaceutical industries? The last thing drug companies in the United States would want was BSE, because bovine by-products are used in a wide variety of pharmaceutical and cosmetic products—everything from insulin to bovine placenta for estrogen in antiwrinkle creams. And if a country finds BSE, they are no longer able to source their products from the native cattle population.

Whatever the reason, we and our sheep were now caught in a political maelstrom.

8

QUARANTINE

uarantine papers! Wayne and Sam just gave us quarantine papers!"
Marybeth Whitten shouted over the phone.

"Slow down, Marybeth, slow down. What do you mean 'quarantine' papers?" I asked. We had agreed to the voluntary quarantine repeatedly and had cooperated with the USDA. Why would they now put us under an official quarantine?

"They gave us a paper that says we are under formal quarantine, and when I asked them what this was all about they wouldn't say. Said they couldn't. They threatened they might quarantine the entire farm," Marybeth said.

"Let me make some phone calls, and I'll call you right back."

I quickly explained to Larry what was happening as I dialed Senator Leahy's office. Bob Paquin answered.

"Bob, this is Linda Faillace. Marybeth Whitten said Wayne and Sam just served them with formal quarantine papers."

"What?! Why?"

"They couldn't give a reason why."

"I don't get it. Why would they do that? Everyone agreed to the voluntary quarantine."

"I know. We don't understand either."

Bob sighed, "Let me call Bill Smith, and I'll get back to you."

It was pouring with rain around 1:00 that afternoon when a beige USDA Chevy Tahoe pulled up our drive. It was Wayne Zeilenga.

"I don't want that guy in our house," Larry told me, so we walked out to the driveway. Wayne was alone, and, when he saw the look on Larry's face, he decided it would be safer to remain in his vehicle. He rolled the window down and asked, "How are you doing?"

"What do you want?" Larry growled, knowing full well what was coming.

"I thought it would be better to bring you this in person instead of mailing it," Wayne said, holding papers out the window.

"That's very generous of you," Larry replied sarcastically.

"Look, I'm only doing my job," Wayne retorted, shaking the papers again, which were now getting wet. I walked over and took the papers from him.

"I don't mean this as a threat," he said, "but let me tell you, if you tell anyone about this, you go to the media, or you get greedy and ask for a lot of money, the USDA will put you out of business. And don't think they haven't done it before."

Wayne looked at me—it was safer than looking at Larry who was angrily staring him down. "You know the imported European cattle?" he asked.

I nodded.

"There was a farm down south; I think it was in Alabama." Wayne continued, "The farmer refused to give up his animals. The USDA kept on top of him, even after they took his animals, and they put him out of business."

Wayne glanced at Larry, who continued scowling at him, and then quickly looked back at me. "I'm just trying to be helpful," he said. And with that he rolled up the window and drove away, leaving Larry and me standing in the rain holding the quarantine papers.

The quarantine papers were dated October 8, 1998—the day of our meeting at the Vermont Department of Agriculture. This meant that Detwiler and the others knew they were going to quarantine the farms when they met with us. Why then did they ask us to agree to continue the voluntary quarantine? What about the meeting the following day at Leahy's office with Bob Paquin? Detwiler thanked us for being so cooperative. Was that just a show for Bob Paquin so he could report back to Senator Leahy how helpful the USDA was being? Why was this necessary? It felt like bullying tactics.

As the rain continued, Larry and I sat at our dining room table and read the following, printed on Vermont Department of Agriculture letterhead:

Dear Dr. & Mrs. Faillace:
QUARANTINE ORDER Pursuant to 6 V.S.A. Section

1157, Dr. Lawrence and Linda Faillace RR1 Box 122-3, Warren VT 05674 are hereby notified that all your sheep flock and any goats are under quarantine as of October 8, 1998.

Those animals shall not be removed from the following-described premises: Dr. Lawrence and Linda Faillace RR1 Box 122-3, Warren VT 05674. You *may not* allow sheep, goats or other ruminant animals to be brought into the same premises as the quarantined animals. The animals may not be removed from the premises except under permit issued by the Commissioner or his designee. This quarantine order takes effect immediately and shall remain in effect until lifted by the Commissioner of Agriculture.

This quarantine is necessary because the above-described sheep:

( ) Are infected with a contagious disease;

( ) Have been exposed to a contagious disease;

(X) May be infected with or have been exposed to a contagious disease;

( ) Are suspected of having biological or chemical residues, including antibiotics, endotissues which would cause the carcasses of the animals, if slaughtered, to be adulterated within the meaning of 6 V.S.A. Chapter 204; or

( ) Are owned or controlled by a person who has violated any provision of this part, and the Commissioner finds that a quarantine is necessary to protect the public welfare.

The Commissioner will lift the quarantine when he determines that your sheep are no longer deemed to be at risk of infection for a contagious disease.

This quarantine is subject to the following terms:

Animals may leave the premises only under conditions authorized by the Commissioner or the United States Department of Agriculture.

In the event you wish additional information, you may contact Dr. Russell Laslocky, Director, Animal & Dairy Industries of the Vermont Department of Agriculture, Food

& Markets, 116 State Street, Montpelier, VT 05620-2901. Telephone: 802-828-2426.

You are advised that you may request a hearing on this quarantine order. Your request should be made to the Commissioner of Agriculture at the above address within 15 days of receiving this quarantine order. You are further advised that a request for a hearing does **not** stay the quarantine order and that it remains in full force and effect.

Pursuant to 6 V.S.A. §1157 (f) it is unlawful to violate the terms of this quarantine order. Any person who knowingly violates a quarantine order shall be subject to a fine of not more than $5,000.00, imprisonment of not more than six months, or both. Any person who knowingly violates a quarantine order and causes the spread of a contagious disease beyond the quarantine premises shall be subject to a fine of not more than $15,000.00, imprisonment of not more than two years, or both.

Dated at <u>Montpelier</u>, Vermont this <u>8th</u> day of <u>October 1998.</u>

[Laslocky's signature]
Dr. Russell W. Laslocky, Director
Designee for the Commissioner
I have read and understand the conditions of this quarantine

_____
Name

_____
Date

Needless to say, Larry and I never signed the papers. There was no way we would agree with *May be infected with or have been exposed to a contagious disease* for a disease that our sheep didn't have and, furthermore, that didn't exist in sheep. To add insult to injury, the following day we received an eleven-page letter from the USDA that outlined their position and presented their four "options."

The letter reviewed the BSE situation, Foster's research from 1993 published in *Vet Record* showing experimental transmission of BSE to sheep in the lab, concern that European sheep could have eaten feed contaminated with infected meat and bonemeal (MBM), and the fact that the UK and France had removed high-risk sheep and goat tissues from human and animal food chains. Then the USDA discussed the importations of the sheep: how many animals were imported (sixty-four), which states they went to, and how many were sold to other farms.

The USDA admitted, "After an evaluation of the Belgium surveillance system on an equivalency basis, the animals were granted 'A' status in the United States [scrapie] flock certification program. The significance of this is that they would now be eligible for certified status and be allowed sale throughout the United States without restriction."

The letter continued that "the sheep originated from Belgium, a country which has had five cases of BSE in native cattle and is known to have high-risk factors such as the importation of British MBM."

The United States had the same risk factors and had imported British MBM right through 1997.

"If the sheep were to develop signs of a TSE, we would not be able to differentiate between BSE and scrapie by routine laboratory testing. This differentiation may be made by a series of mouse bio-assays which usually takes at least two years. As a part of the surveillance protocol for the imported sheep, one animal was culled and tissues submitted for diagnostic purposes. Histology findings revealed a diagnosis suggestive of scrapie. All other tests thus far have not found additional evidence of a TSE."

Detwiler had rebuked Leon Graves for claiming Mr. Freeman's sheep might have scrapie or BSE. We saw the test results, and nothing was "suggestive of scrapie." The ewe had severe mastitis. The USDA told us and Senator Leahy's aide one thing and said something different in their memorandums.

The letter then listed the options and the resulting "pros" and "cons." Option 1 was to buy all the animals at "fair market value."

Pros

- With the slightest possibility that the sheep are infected with a TSE agent unknown to the United States this action would protect both human and animal health by preventing the entrance of tissues into both the food chains and eliminating the possibility for lateral spread.
- If the sheep had remained in Europe and would be considered exposed, certain tissues would be prohibited from

the human and animal food chains. This proposed action would go beyond the European precautions and would help maintain consumer and trading partner confidence.

Cons

- The animals may not be infected with a TSE. This may destroy new genetics imported into the United States to improve our domestic sheep industry.
- If fair market value were paid, the cost may be considerable as reported by the importer. The Agency may have to justify paying more for the sheep than the imported UK cattle.

Option 2 involved purchasing the flocks for research purposes, keeping the sheep alive, and saving the genetics.

Pros

- The sheep would be maintained and monitored (by sampling as above) and prohibited from the human and animal food chains.
- This option may be more palatable to the owners as there is the possibility of preserving the genetics.
- Given the possibility that the sheep may not have been infected with a TSE, the flocks would not be destroyed unnecessarily.

Cons

- This option will be costly. One might ask the question: Should the United States Taxpayer shoulder the bill to preserve the genetics for two flocks?
- If the sheep are infected and they are not housed in a higher security location, additional premises may be contaminated with an agent unknown to the United States.

Option 3 involved putting the animals under a state quarantine, purchasing any that would be culled for slaughter, and collecting samples.

Pros

- If all the original imports and culled progeny are eventually sampled and no additional evidence of disease is found, the genetics would be preserved.

• This may be more palatable for the owners than the above options.

Cons

• A legal quarantine does not absolutely guarantee that these sheep would stay on the premises. Movement off of the premises of any sheep may reduce consumer and trading partner confidence in our ability to restrict this flock.

• If the sheep are infected with a TSE agent (unknown in the United States) that possibly may have human health implications, there may be a risk for the caretakers and accredited veterinarians.

• With no additional evidence, the State may not be able to hold the animals under a quarantine.

• This will save the Agency the initial funding for purchasing the sheep outright. With multiple trips for necropsies and payment for disposal in an increasing flock, it may not save money over the long run.

Option 4 was to "maintain the flocks under the current voluntary protocol for purchasing, sampling, and disposing of culls."

Pros

• May be the most palatable to owners.

• With current evidence may be the only option for the State without accruing legal liabilities.

Cons

• If the sheep are infected with a TSE unknown to the United States this may allow the spread of infection.

• No legal measures to prevent movement of sheep into the human or animal food chain.

• If we find a problem in future years, the number of animals to be destroyed will be significantly larger.

• Without restrictions, we will most likely lose consumer and trading partner confidence.

The letter concluded with:

> It is recommended that the U.S. Department of Agriculture
> (USDA) Animal and Plant Health Inspection Service (APHIS)
> choose Option 1 to provide funding for the purchase of the
> imports and their progeny to prevent any risk of an introduction
> and subsequent spread of BSE into the United States. USDA,
> APHIS has had actions in place to prevent an introduction of
> BSE for well over 10 years. We should continue with all meas-
> ures which would continue to protect us from an introduction of
> the disease.

Robin Aldrich called us when the USDA served them with quarantine papers. "What are we going to do?" she cried. Because Robin had
a ewe from the Freeman's farm, her entire flock of twenty-one sheep
was at risk. Robin and her husband, Barry, had purchased a beautiful
piece of land in Vermont's Northeast Kingdom, adopted four foster
children, and hoped to eventually become full-time dairy sheep
farmers with some American sheep, the sheep from Mr. Freeman, and
a New Zealand ewe they bought from us.

"I don't know," I replied, frustrated. "There's no science to back up
their actions."

"So how can they get away with this?"

"They shouldn't be able to."

"Are you going to fight the quarantine? We are."

"Larry and I are still talking about all possibilities." I was tormented.
Our customers, who put their faith in us, who wanted to be part of the
nascent sheep dairy industry, were now being harassed by the USDA.
And there was nothing I could say to reassure them. I promised Robin
I would keep her apprised of the situation and told her we would keep
pushing for the science. But my hopes were wearing thin.

I never imagined our family would one day be embroiled in an
international political situation. It was obvious from reading the
options letter that science did not matter to the USDA, only public
perception and bureaucrats' jobs. The USDA was using fear to justify
their means. Under option 3 they mention that caretakers' and veteri-

narians' health could be at risk from handling the animals. Yet TSEs are not spread by contact. No one's health was at risk from interacting with the animals.

And I was surprised by the number of references to "consumer and trading partner confidence." What did our flock of sheep have to do with international trade? BSE did not exist in sheep, so why would U.S. markets be concerned?

On Election Day 1998, Larry, Mr. Freeman, and Tom Amidon met with Senator Leahy, who was very friendly and listened as Larry updated him on the science concerning the issue. "It doesn't make sense," he said. "I'll see what I can do."

A few weeks later Larry and Mr. Freeman met with Governor Howard Dean. "Dean and his wife are medical doctors, and they should be able to see that the USDA is not using science," Larry told Mr. Freeman as they walked through the lobby of the Vermont state house.

Quietly they were ushered into the governor's office. When Dean rose from his desk, he was stiff and formal, unlike Senator Leahy. Not a good sign.

Again, Mr. Freeman and Larry reviewed the situation for him and asked for help. "There's not a lot I can do," Dean replied. "This is complicated. More scientific than I can understand."

"But no sheep in the world has ever gotten BSE," Mr. Freeman said.

"I'm sorry," Dean replied. "I tend to agree with the USDA and err on the side of caution."

Meanwhile, the forces behind the USDA were gaining steam. Despite the fact Detwiler repeatedly told us to "keep quiet" and "not go the media," the November 13, 1998, issue of the American Sheep Industry Association (ASI) newsletter had the following announcement:

> State and federal officials have been working for many years to prevent Bovine Spongiform Encephalopathy (BSE) from entering the United States. Those efforts were recently upgraded to include the quarantine of a small number of sheep of Belgium [sic] origin and their offspring.
>
> The original sheep entered the United States prior to a 1996 ban on importing sheep before Belgium reported BSE

in cattle and before the European Union reported a risk of BSE existing in its population. The flocks, located in the Northeastern United States, have been under federal and state surveillance since that time.

American Sheep Industry Association leaders support the efforts of government officials to combat BSE. "We're all being extremely conservative about it," said Paul Rodgers, ASI's Director of Research and Education. "There has never been a single case of BSE in this country, and we'd like to keep it that way."

Reading the article made me angry. Paul Rodgers was the director of research and education for the ASI. Shouldn't he be talking about the fact that no sheep in the world had ever contracted BSE? Instead readers were left with the impression that sheep in the EU developed BSE and getting rid of ours was a necessary evil.

In January of 1999 ASI passed a directive trying to force us to sell our sheep to the USDA: "Be it directed that ASI urges USDA/APHIS to: Sunset any indemnity offer to owners of imported Belgian sheep and offspring by 6/1999 with a limit on the numbers of sheep to be paid for, and affect the expeditious depopulation of the original imported animals and all offspring."

Vermont's ASI representative, Phil Hobbie, met with us and explained how he was able to persuade ASI board members to extend the indemnity period from March to June. I questioned Phil, "What do you mean they will *extend* the indemnity period? What does ASI have to do with paying us? I thought this was between us and the USDA." Besides, Larry and I had no intention of selling our sheep. I was determined that science would win this battle.

"Well, you have to understand," Phil replied, sounding annoyed, "some people at ASI are not happy with you guys and don't think you should even be compensated." Phil and his wife raised wool sheep in southern Vermont. During the day he was employed at a local university, and he tended his sheep at nights and on the weekends, as did most Vermont sheep farmers—very few were full-time shepherds. Phil worked hard for every penny he received from his sheep, and with wool prices hitting an all-time low, he was grateful for anything he received.

Rumors were flying about how much money the USDA offered us, sums of millions of dollars, but all were rumors. Phil explained how he "went to bat for us," yet we didn't seem appreciative. Why couldn't we just sell our sheep and be done with it?

Tom Amidon decided to ask Dr. Gerald Wells of Veterinary Laboratories Agency in Weybridge, England, to look at the test results from three additional ewes that were culled from Mr. Freeman's Skunk Hollow Farm due to mastitis and write his scientific opinion. Wells reviewed the histopathologies, which noted some astrocytosis, and he claimed:

> This glial cell response occurs in many neurological disorders and cannot, therefore, be regarded as specific. Nevertheless, its occurrence in these cases suggests that sheep from this farm may have a neurological disease, albeit not specifically identifiable as scrapie, and not necessarily manifest as a clinical disorder.
>
> This raises certain concerns, some of which may be relevant to the occurrence of atypical expression of transmissible spongiform encephalopathies or prion diseases.

I read this with bewilderment. The astrocytes response was completely normal and expected in animals with mastitis. There were no necrotic neurons and no spongiosis, yet Wells steered toward a diagnosis of an "atypical TSE."

The letter continued, saying that even though the animals were not showing clinical signs of scrapie, they were being tested for the disease.

> In the sheep examined from Skunk Hollow Farm there has been a consistent failure to be able to demonstrate the accumulation of PrP [prion-related protein] in the brain. Thus in summary, there is no conclusive evidence from the material and reports to which I have had access that sheep from this flock are suffering from scrapie.

"Okay, finally," I thought. "He admits there is nothing wrong." But no, wait. He continued:

In as much as evidence of a glial reaction has been demonstrated in these cases there is the suggestion that these sheep are, however, suffering from a neurological disorder. This may or may not be a contagious disorder.

He concluded his letter with:

… the obvious concerns arising that we know less about BSE in sheep than we know about scrapie and that BSE may not be distinguishable from scrapie even based on a range of tests including the mouse bioassay. Given the source of these sheep the concern that they may have been exposed to meat and bonemeal containing the BSE agent cannot be refuted with certainty. The option one proposed by the USDA would, therefore, seem the most responsible under the circumstances.

But in fact our sheep's exposure to meat and bonemeal *could* be refuted with certainty. We had the feed records from all the flocks and the feed mills, and they were certified by the Belgian government. And how did Wells know about the USDA's letter with the options?

The last two pages contained his lab notes. Here he was required to be scientific and not allow his political persuasions to interfere. He reviewed the written material he received, looked at the sheep tissue under the microscope, and ran another test (an immunohistochemical assay) to search for abnormal PrP. His remarks were:

The absence of convincing vacuolar changes in the brain of the sheep prevents the confirmation of a spongiform encephalopathy which is considered the hallmark of clinical cases of scrapie.

The prominence of the astrocytic change in this brain is, nevertheless, suggestive of an underlying neurodegenerative process, although without other pathological changes this can not be regarded as in any way diagnostically specific. Similarly, the changes described in the cuneate nuclei are also non-specific and, in fact, commonly present in adult sheep.

"Non-specific and commonly present in adult sheep" was not the conclusion he chose to reiterate in his letter, but instead he noted that he was in contact with the USDA and was providing them with a second opinion. This letter was the spark that started the fire.

9

POLITICAL PRESSURE

One by one our customers sold their animals to the USDA. The USDA paid them the original purchase price plus a premium of up to 40 percent—$6,000–$7,500 per animal—and allowed them to keep the offspring and any semen they had collected. But people were not sure who or what to believe. Were Larry, Mr. Freeman, and I right that there was nothing wrong with the sheep and that this was a political battle in which our sheep were pawns? Or was the USDA correct that the sheep might harbor some unknown disease that could possibly fatally infect them and their families?

One customer had a freezer full of lamb but wouldn't eat it. Others were worried the USDA might come back and try to take additional animals on their farms. And Wayne Zeilenga fanned the fire by telling the Whittens that they should sue Mr. Freeman because they "probably had nv[new variant]CJD." Fear was running rampant.

Larry and I figured the best way to end this nightmare was to meet with Detwiler's bosses and demonstrate that our sheep were healthy and that no sheep in the world had ever naturally contracted BSE. We flew over three European experts: Dr. Bernard Carton; Dr. Piet Vellema, a sheep and scrapie expert from the Netherlands; and Professor Emmanuel Vanopdenbosch, the current head of SEAC. Dr. Vellema was quite tall and slender with short red hair, while Professor Vanopdenbosch was slight with dark hair and dark eyes, and was very soft-spoken. As we drove to the hotel in downtown Washington, D.C., the scientists asked about our situation and listened quietly while we explained everything: from the importation, to Detwiler coming to our farm, to the fact that politicians consistently avoided us.

"What can we do?" Professor Vanopdenbosch asked.

"Well," Larry began. "Detwiler is in charge of keeping BSE out of the United States—"

"Yes, I know Dr. Detwiler," Professor Vanopdenbosch interjected.

"She is on the same committee I am . . . as an invited guest," he added.

"She appears to be the driving force behind wanting to kill our sheep," Larry continued. "The reason for inviting all of you to this meeting is the hope that we can address the issue with science, not politics. If we can get to Detwiler's superiors and show that our sheep are not a risk, then this whole issue can be resolved."

The next morning was a gorgeous spring day, and I was optimistic. I called Mr. Freeman and told him everyone was there and I thought the meeting would go well. Mr. Freeman laughed. He liked optimism, and I had plenty of it. Mr. Freeman had tried repeatedly to find a compromise, but the USDA had refused all his offers so far.

"If *you* get the USDA to work with us, I'll reimburse you for all your expenses for putting on this meeting," he teased.

"That's a deal!" I said, smiling.

The USDA is located in the Jamie L. Whitten building—a tall, gray, formidable-looking structure that was a stark contrast to the beauty of the day around us. There were no gracious gardens, nothing to soften the cold, hard exterior. I felt my stomach tighten. We signed in with the receptionist and were escorted to the third floor. The halls were barren, painted hospital green, and many of the doors had glazing over the windows to prevent others from seeing in. No one we passed was smiling. "What a horrible atmosphere to work in," I thought. I noticed the clicking of my heels echoing down the corridor and tried to walk more quietly.

The escort took us to a small room that was overpowered by a large rectangular table surrounded by sixteen chairs. No one was there. Bob Paquin said that Senator Leahy and Michael Dunn would come for the beginning of the meeting, and Ed Barron and Michelle Barrett from Leahy's office would be there for the entire meeting. But besides Detwiler and a representative from the National Grange, we did not know who else would come. Our hope was to go straight to the top and that Dan Glickman, the USDA secretary of agriculture, would also make an appearance.

We waited almost fifteen minutes. "Not a good sign," I thought. "They are using the intimidation tactic. Make your opponent sweat." Finally the door opened and a small crowd filed in. I quickly scanned the group. Senator Leahy was not among them, and neither was Ed Barron,

Michael Dunn, or Secretary Glickman. My heart sank. We had paid to fly the Europeans over, yet no one with clout would be at the meeting.

Everyone nodded heads, said hello, then settled around the table: six of us at one end, and seven of them at the other. Larry and I handed out the portfolios: a three-ring binder filled with articles, photographs, and the history of our company—a demonstration of all the love, thought, and effort that went into our business. Michelle Barrett handled the introductions. From her right side, seated around the table were Courtney Billet, another aide to Senator Leahy; Louise Calderwood of the Vermont Department of Agriculture; Linda Detwiler; Dr. Alfonso Torres, who was deputy administrator of veterinary services of USDA/APHIS and Detwiler's boss; my father, Dr. Glenn Cahilly; Professor Emmanuel Vanopdenbosch; Dr. Piet Vellema; Dr. Bernard Carton; Larry; myself; Dr. Joe Gibbs of the National Institute of Health and a TSE expert; and Dr. Tom Walton, another USDA employee.

After the introductions, Michelle motioned to Larry. "Dr. Faillace, would you like to take over from here?" Larry nodded and stood. "Thank you all for meeting here today. As you are aware, the USDA is concerned our sheep could be susceptible to BSE. Obviously, Linda and I vehemently disagree and have arranged for Drs. Carton, Vellema, and Vanopdenbosch to present information and answer any questions you may have. We would like to see this entire matter resolved as soon as possible so we can get back to farming.

"Professor Vanopdenbosch, would you like to start?" Larry asked, as he sat down. Professor Vanopdenbosch nodded and proceeded to explain the purpose of the European committee of which he and Detwiler were members. He spoke of how there was concern in Europe that the sheep population could have been exposed to BSE. But, he emphasized, no sheep outside of laboratory experiments had ever contracted BSE, and there was not an increase in the incidence of scrapie when there was an increase in BSE in cattle. Therefore no action was being taken against the European sheep industry in the interest of public health.

Dr. Vellema then spoke about the scrapie program and explained how all the imported sheep had been monitored for scrapie. None of the flocks had ever had scrapie; in fact, the East Friesian breed of sheep had never had a documented case of scrapie.

Dr. Carton continued by explaining how he was very familiar with all the shepherds, was the veterinarian for most of the flocks, and knew what all the animals were fed. None were ever fed meat and bonemeal. Dr. Carton had received signed statements from the shepherds and the feed mills and had all the documents certified by the Belgian Ministry of Agriculture. He offered copies to everyone at the meeting. They all politely refused.

When Dr. Carton finished speaking, I felt a strong sense of satisfaction. The Europeans had done an excellent job, and now we could get this whole thing over with. Larry thanked our guests and opened the meeting for questions.

Silence.

My heart was pounding. "Why don't they have any questions? For the past forty-five minutes they listened as the Europeans spoke. Surely they want more details," I thought. I looked at their faces, one by one. Most were staring silently at the table.

Finally Dr. Torres spoke. He looked compassionately at me. "I'm sorry, but we are under political pressure, and you will have to surrender your sheep."

"Political pressure?! From whom?" I asked. I was tired of this excuse.

"We can't say."

"You mean to tell me you are going to kill our sheep even though nothing is wrong with them?"

Larry leaned over to hold my hand. He could see the tears welling up in my eyes. "It's important to stay calm and in control," he had told me before the meeting. He squeezed my hand to remind me.

Dr. Torres replied, "It's a risk we can't take."

"All the information you are basing your decision on was known since 1993—three years before we imported the sheep. Yet you approved our importation. Why?"

Again it was silent. Dr. Torres finally spoke, "Well, I was not around when that decision was made." Everyone else nodded their heads in agreement. None of them wanted to implicate themselves, especially Detwiler. I knew she was a USDA employee in 1996.

"We will pay you fair market value," Dr. Torres said, trying to soften the glares that they were all now receiving from Larry and me.

"I am sick of hearing about 'fair market value,'" Larry retorted. "There is nothing wrong with our sheep, and we should not have to sell them."

"What about keeping the sheep quarantined and using the tonsil test?" Professor Vanopdenbosch suggested.

"That sounds like a good idea," Dr. Joe Gibbs said, nodding his head in agreement. "What do you think?" he asked Dr. Torres.

"Well . . ."

Louise Calderwood jumped in, "We should *not* risk keeping those sheep around! There are only a dozen or so families involved, and we should get rid of those sheep and get the whole thing over with!"

Who was this woman? I had never met her before, and here she was as a representative for the Vermont Department of Agriculture. And why was she being so emphatic? Detwiler gave Louise a slight smile—a comrade.

"I'm sorry," Dr. Torres said again, looking directly at me. "We will pay you fair market value."

"Just what is fair market value?" I demanded.

"Well, what do *you* think your animals are worth?" he asked, turning the question back on me.

"If you look at our business plan, we would have earned $11.3 million dollars over the next ten years," I answered without skipping a beat. Larry and I had spent the past five years working on the business plan, and I could rattle off any number that was in it.

Dr. Torres's and Detwiler's mouths dropped open. "I don't know where you are getting your numbers from," sneered Detwiler. "I haven't seen any invoices that would support what you are saying."

"You have to understand that the protocol for what the USDA pays in instances like this has been in place for over a hundred years," Dr. Torres added. "If you want more than market price plus a genetics component for your animals, you will have to take a legal approach."

Detwiler was still angry. "The USDA has not been stingy with money when it came time to buy the imported sheep we've bought so far." A few seconds later she turned to Michelle Barrett and asked, "Does Senator Leahy's office have access to any money that can be used?"

"I can look into it," Michelle replied.

"But I don't want to sell our sheep!" I cried as the futileness of our situation set in. Mr. Freeman was right, the USDA would not shift.

"So what exactly is 'fair market value'?" Larry demanded, as he put his hand protectively over mine.

"Well, um," Dr. Torres stammered. He looked at Detwiler for suggestions. She said nothing. "Why don't we talk about it and get back to you," he said meekly.

"We've been hearing this for the past seven months!" Larry angrily exclaimed.

"So you're telling us that because of political pressure there's no way we can save our sheep?" I asked, still in disbelief.

Everyone stared at the table. The Europeans sat quietly, watching.

"Well if you are going to pay 'fair market value,' then I want to know exactly what that figure is." Larry said irritably.

Sensing a weakening, a look of relief crossed over Dr. Torres's face. He turned to Detwiler. "Do you think we could have this ready in the next week?"

"I don't see why not," she replied.

Dr. Torres again looked at Larry and me sympathetically. "We will send you a letter next week explaining our offer."

I sat there, numb. "This is not how the government is supposed to work," I thought. "Who's controlling whom? How is it that commercial interests can force a government agency to do their bidding? There is nothing wrong with our sheep; the USDA agrees with us, yet they want to kill them, pay us off, and be done with it."

I thought of Mrs. Friendly with her smiling face, looking up at me. Fromme, BB, and the rams with all their unique calls. Even Kiwi, our New Zealand ram, had a different voice than the rest, and Francis could mimic all of them perfectly. We were responsible for all of their lives. We had brought them all the way from Belgium, the Netherlands, and New Zealand. We helped them have their young, raised them in the best way possible, provided the best pasture, and loved every one of them. Yet here was an organization whose mission and entire reason for existence was to support American agriculture, telling us we had to kill our animals. I could feel the tears welling up again.

"What about sending the animals back to Belgium?" I asked.

"Well, I guess we could consider that," Dr. Torres said quietly. He turned to the Europeans. "What do you think? Could the animals go back?"

The three men looked at each other. "It would be up to our ministry," Professor Vanopdenbosch replied.

Dr. Torres turned to Detwiler. "Look into that, would you?"

Detwiler grimaced.

• • •

When the meeting ended, Dr. Gibbs stood and turned to me. "I'm sorry," he said in a gentle voice. "I wouldn't hesitate to eat a leg of your lamb. But you have to understand that if the United States was *perceived* to have BSE, it would cause the stock market to crash."

And then he hugged me. All I could do was lean against him and cry. We just wanted to have a family farm, raise our animals, and help the sheep industry. Why were our sheep so important to the stock market? Millions of American sheep were not even monitored for scrapie or any other TSE, yet the USDA insisted our small flocks could make the entire American economy go into a tailspin—for a disease that didn't even exist in sheep.

The somber mood in the air made it feel like a funeral. Dr. Torres now stood beside the two of us. "I'm sorry, Mrs. Faillace," he said, offering condolences, putting a hand on my shoulder.

Detwiler and Louise Calderwood kept their distance. Senator Leahy's aides shook our hands. This was not the outcome I expected.

Everyone was silent until we were outside. The bright sunshine was in sharp contrast to my mood, and I was deep in thought when Larry finally spoke. "Did you know Louise and I were at Virginia Tech at the same time when I was working on my PhD?"

"I hope she wasn't an old girlfriend," Dr. Carton joked. Everyone laughed, wanting to relieve some of the pressure.

"That was unbelievable," I said. "At least Dr. Torres and Dr. Gibbs were sympathetic."

"I don't think Dr. Torres was sincere," Larry countered.

I sent a nasty look his way. "I do," I insisted, refusing to believe Larry's cynicism. I always tried to find the best in people, and to me Dr. Torres seemed concerned. I desperately wanted to believe that someone in the USDA cared.

The flights back to Europe were not until the following day so we used the remaining time to explore Washington, D.C. We walked for hours—to the Washington Monument, the Jefferson Memorial, and then the Lincoln Memorial. As we stood on the steps of the Lincoln Memorial, a group of men with name tags were gathering for a photo.

"Would you like me to take your photo for you?" I asked.

"That would be wonderful! Thank you, ma'am," replied a tall man wearing a large cowboy hat. I took their camera and began adjusting the lens.

"You just push the button on the right," he yelled over.

As I focused, I noticed their badges, "National Cattlemen's Beef Association."

"What are the odds?" I thought, shaking my head. I took a few photos and handed the camera back. While the rest of our group was on the steps, Carton stood beside me the entire time and saw the name tags also.

"Thank you very much. I sure do appreciate it, ma'am," the cattleman said, tipping his hat toward me. "Would y'all like me to take your photo, too?" he motioned to the camera I was carrying.

"That would be very nice," Carton replied, taking the camera from me and handing it to the man. He pulled me over to the others. "Gather for a photo," he told them. As we stood there smiling, Carton put his arm around me and squeezed. The irony of the situation was not lost on him.

10

WARREN MEADOWS

A week passed, then two, then three, yet still no letter from the USDA. In week four it arrived and read: "We will pay you *fair market value* . . ." No figure. No defining of "fair market value." The letter rehashed the four options and reiterated that the USDA wanted to go with option number one.

Larry and I were fed up. We dug in our heels. There was no way we would give up our sheep for an unspecified amount for a disease that didn't exist. It was time to get back to farming. If the USDA wanted the sheep they would have to find another way because they were not for sale. We would stand by our principles, our morals, our family, and our flock.

Larry and I were very active with the VSBA and the Vermont Cheese Council. The president of the American Cheese Society (ACS) contacted me about the upcoming annual conference being held in Shelburne, Vermont.

"Can you organize the Sunday tour?" she asked. Every year the conference was held in a different venue around the country and ran from Thursday through Sunday. The Sunday tour was a chance to see local cheesemaking operations.

"Sure!" I told her, and I began listing possible places to visit.

"And don't forget your place," she added. "Lots of people are asking to see what you guys are doing. Your cheesemaking building sounds great!"

"But . . . there are plenty of other farms . . ." I stumbled.

"Yeah, but a lot of people have already seen them. People are excited over what you are doing and want to see your place, so make sure you include yourselves in the tour."

"Oh, no," I thought. Here was something I normally would have been thrilled over, but now we would have cheesemakers, chefs, distributors, and food writers visiting us, asking about our operation, all while our sheep were under quarantine and we were forced to remain quiet.

Plus we were still building the cheese facility. From dawn to dusk the entire family worked together: sawing, nailing, staining clapboards, installing windows and doors. The children were incredibly responsible and very adept at taking measurements and using power tools. I think this is true of many farm families, because everyone has to work together for a living. We wanted a bright, airy, affordable building that would be large enough to host our cheesemaking courses and decided to use a solar-style building with breathable fabric on the sides and exhaust fans on either end to ensure a good air flow on warm summer days.

I was staining clapboards in the visitor's center on a summer's day when I noticed Jackie going back and forth between Heather and Francis, whispering and writing something on a piece of paper. This continued throughout the afternoon. When we cleaned up the tools that evening, the children asked us to sit with them at the picnic table under the large maple tree. After a hard day's work it was nice to take a break. "We want you to go to Montreal for your anniversary," Heather told us. Larry's and my wedding anniversary is July 30, and each year we took a trip to celebrate, but this year we would not be able to give up the time.

"That's really sweet of you," I said, "but we have too much work to get done." The ACS conference was the following weekend.

Jackie pulled out the piece of paper and laid it in the middle of the picnic table. "We've made a list of all the jobs left to finish and divided them up."

Larry and I looked at the list:

- Finish cutting clapboards
- Caulk around floor edges
- Install dartboard
- Screw clapboards
- Finish staining clapboards in visitors' center
- Stain outside clapboards
- Epoxy floor

He laughed. "There's no way we can leave you all here to do this while your mother and I take off for Montreal! Some of these jobs involve using the saw, the jigsaw, and the nail gun. That would be irresponsible of us to leave children alone using power tools."

Francis bristled, "We are not 'children' anymore!" He was a mature fifteen-year-old.

"Yeah, we can handle this," Heather chimed in.

Larry smiled. "I tell you what," he said. "How about if we all work together extra hard and try to complete these jobs? If we get them done in time your mother and I will go to Montreal."

For two days the children worked with us from 6:00 in the morning until 1:00 the following morning. It was our best anniversary present ever.

The weather was beautiful for the entire American Cheese Society conference. The Sunday tour started in Burlington, and we traveled by bus to Willow Smart's farm to see her new aging cave. As the designated tour guide, I explained farming and cheesemaking in Vermont, how all the cheesemakers work together on the Vermont Cheese Council, and how people like Willow were milking sheep and making wonderful cheeses that sold so quickly she couldn't keep up with the demand. Next it was off to our farm. As the bus passed through Warren Village and began the ascent up Brook Road, I told the group about Rootswork and Anne Burling. A former Vermont state representative, Anne had purchased a ninety-two acre property in Warren in the mid-1980s with the vision of having organic farmers work the land together. In 1995, together with a few other enthusiastic community members, Anne formed Rootswork—a nonprofit organization dedicated to promoting and fostering community-centered organic and sustainable agriculture. Anne urged Larry and me to become Rootswork members and offered her land for grazing our sheep. We eagerly joined Rootswork and soon built a solar barn and the cheese facility on land we leased from Anne.

Abutting Anne's property was a large, two-story schoolhouse owned by the town. The building was a school from 1897 to 1974, and then housed a variety of businesses, with its last incarnation in the early 1990s being a small convenience store. By 1998, the building was run-down, had structural problems, and needed a new heating system—something the town was not willing to invest in, so it was abandoned.

A group of us met with the Warren town selectboard to discuss the possibility of Rootswork leasing the schoolhouse. The selectboard agreed, happy to have the building put to use. They rented it with the caveat that it was up to us to fix the building; the town was not willing to spend any money on it. And if we wanted to maintain the grandfather clause for a retail establishment we had to sell at least one item every year. Over the next few years we installed new plumbing, new electricity, and painted the inside. The town was so thrilled with the improvements that they shared the cost with us of a new heating system and a new standing-seam roof.

Some Rootswork members talked about reopening the store. After a year of discussions, Bruce Fowler, a landowner whose property was adjacent to Anne Burling's, decided he would take matters into his own hands. While Larry, the children, and I were busy getting the cheese facility ready, Bruce worked inside the schoolhouse converting the first-floor room into a store. It was in a shambles, filled with years' worth of junk and trash, while a large cable held up one wall.

As the ACS tour bus climbed the last hill toward the farm, I told the group that I had yet to see Bruce's work and encouraged them to give him a kind word. Today would be the first day the store was open. The bus pulled into the large parking lot, and everyone made their way to the cheese facility where Jackie and Larry were waiting to give the tour. I stayed in the back of the group and couldn't help but grin ear to ear as I listened to my twelve-year-old answer questions on equipment, starter cultures, and the process of creating artisanal cheeses.

People wandered through the beautiful gardens and gradually made their way into the store. Bruce had set up one aisle with a few groceries, mostly canned goods, and an old three-door cooler with a few half-gallons of milk and some soda. Roger Hussey, a close friend and fellow Rootswork member, wore a straw hat and played the banjo beside a gingham-covered table ready with chess and checkers. A little country ambience.

The sun was now high in the sky and people began to move more slowly. Time for lunch. So we boarded the bus and drove six miles north to the Round Barn Farm. Formerly a dairy farm with a large round barn, it became a luxurious country inn when the Simko family bought and renovated it. Cooking from the Heart was a local catering company that hosted events at the Round Barn and established an excellent reputation for using the best local organic ingredients. They crafted an awe-inspiring feast that afternoon featuring local cheeses, fresh organic vegetables, trout, salmon, and the famous American Flatbread.

George Schenk, founder and owner of American Flatbread, had built a clay pizza oven behind the Round Barn. Before lunch, George explained to the group how his flatbreads were made and the importance of cooking with love—all the ingredients should be raised and harvested with appreciation and the food prepared with love in order to be beneficial and healing to our bodies. As George spoke, Hanna, his twelve-year-old daughter and Jackie's best friend since first grade, prepared the flatbreads. At the end of the meal no one wanted to leave. Everyone was content to stay, savoring the sun-drenched afternoon in this magical place. The tour was deemed a huge success.

On the Saturday night of the American Cheese Society conference the "Festival of the Cheeses" took place. Anyone who wanted to have their cheeses judged was required to send ten pounds of each cheese. The judges sampled all the cheeses and the remainder was displayed for the conference attendees to try. As you might expect, this was the highlight of the festival—all the cheese you could eat.

Even though hundreds of people attended the event, there was still an abundance of cheese left at the end of the evening that anyone could take home. At previous conferences I was able to procure a seven-year-old cheddar, some three-year-old Goudas, and numerous other mouth-watering cheeses. Most people took a few pounds, just what could fit in their luggage.

"Do you have a cooler you might be able to fit some cheeses in?" one of our cheese course students asked Larry.

"How many pounds do you have?" He had seen her collecting cheeses by the armful.

"Oh, I'd say about four hundred pounds."

"Four hundred pounds?!" Larry exclaimed. Then he thought of the near-empty cooler in the schoolhouse. "Yes, we can store it for you."

When the student arrived, Larry, Jackie, and I helped carry all the cheeses to the cooler. "Feel free to take as much as you like," she told us.

It was our third year of teaching cheesemaking. In 1997 we had arranged for Freddie Michiels of Belgium to teach the Whittens and ourselves to make a variety of cheeses. Every year Freddie came back to help teach the courses with Larry and Jackie. We limited our cheesemaking courses to no more than ten students in order for each individual to get plenty of hands-on experience, and this year all three classes were full.

The first day of the cheese class was an unusually hot August day for Vermont and the students needed something to cool down with. "Can you help me get some bottles of water and cans of soda?" I asked Jackie.

As we walked into the store, a BMW with Massachusetts plates cruised into the parking lot and out stepped a nicely dressed couple in their early fifties. "Is there a store here?" they asked.

"Yes, come on in," Jackie answered as they followed us.

"What an incredible selection of cheeses you have!" the woman exclaimed when she eyed the cooler.

Jackie's face lit up. "Would you like to try some?" she offered.

"We'd love to."

For the past two years Jackie had helped set up the Festival of the Cheeses and was familiar with each cheese. Jackie cut samples and described the various cheeses: where they were from, what awards they had won, what the cheesemakers were like, and which cheeses were her favorites. The couple happily purchased $45 worth of cheese.

When Bruce came to the store that afternoon, I pulled him aside and told him, "You have a business partner." He smiled, nodded his head, and said, "Okay."

It was good to have other things to concentrate on besides battling the USDA. Francis was busy with the pastures and keeping the sheep in tip-top shape, Heather milked the sheep twice a day, and Larry and Jackie were in full swing making cheese. Bruce and I talked about our visions for the store. We agreed not to sell cigarettes and lottery tickets. Bruce saw the store maintaining the convenience items it had previously sold, while I envisioned something more gourmet. We both agreed to sell as many local products as possible.

Over the next few years we added more and more products: local Vermont jams and jellies; maple syrup; baking mixes; honey; artisanal breads; homemade pies and cookies; fresh fruits, vegetables, and flowers; locally grown, all-natural meat; locally produced wine and beer; and, of course, Vermont cheeses. All the cheese Larry and Jackie could produce, I could sell in the store. So we added more Vermont cheeses to the repertoire.

An article in the *Times Argus*, a local paper, that fall featured our story. On the cover was a photo of Heather surrounded by her elegant, long-eared East Friesians. Mary Gow wrote:

> In the middle of an East Warren meadow, Heather and Jackie Faillace stand at the end of their green and white mobile milking parlor. Nearby, their brother, Francis, eases open a paddock gate. Heather starts to ring a small hand bell. At the sound of the bell, a bleating blur of white fleece and pink ears charges through the gate and up the wooden ramp into the parlor.
>
> It's sheep milking time for the Faillaces.
>
> Francis, Heather, and Jackie Faillace raise East Friesian dairy sheep and make cheese from their milk. This week, Warren Meadows Cheese, their herb-coated sheep's milk cheese, will go on sale for the first time and will carry the label Three Shepherds of the Mad River Valley. Its sale starts a new phase in their family dairy sheep business. It also marks the commercial debut of Vermont's youngest cheesemaker, twelve-year-old Jackie Faillace.
>
> Francis Faillace's business card reads "Pasture Manager/Sheep Specialist." This fifteen-year-old sophomore routinely moves fences, herds sheep, trims hooves, and tags lambs.
>
> "The milking ewes get a fresh paddock every twelve hours," Francis explains. "The lambs move every two days and the rams every two or three days." Before school every morning and again in the evening, Francis moves flexible fences to set up new paddocks and then herds the sheep into them.
>
> This summer, the abundance of sunny days posed some problems for the sheep. East Friesians are an elegant breed with long legs, very white fleece, and pink ears. With so much sun, their ears and udders risked sunburn. Francis set up a small party-style canopy for them to get shelter.
>
> "But," he explains, "you don't want to give them too much

shade or they'll get lazy and stop grazing. You want them to keep grazing to keep up milk production."

Heather Faillace, thirteen, is the family's sheep milker, and the sheep know it. When Heather goes near the flock, they all gather around her. She scratches their ears, talks to them, tickles their tails, and knows every ewe's idiosyncrasies.

"This is Kiki. Wait until you see her kick in the parlor," says Heather, scratching one ewe behind the ears. "Martha, this one, didn't lamb, but she likes to come along with the rest."

From early summer when the lambs are weaned, until mid-October, Heather milks the ewes twice a day. The entire milking parlor is in an open-sided trailer that can be towed to the sheep.

At milking time, Heather hauls a vacuum milking machine with belly buckets along the line of ewes. Little tails twitch as each ewe has her turn. Kiki kicks so much that Jackie helps hold her to shield Heather from flying feet.

Jackie Faillace started making cheese when she was in the fourth grade. She took her first cheesemaking class and started working with her father on different recipes.

"We made thirteen varieties the first year," she remembers. "The second year was fifteen." So far, she has taken nine courses and spent nearly a thousand hours making cheese and refining recipes.

In the family's solar-style cheesemaking facility, sunlight streams through the white shaded roof and reflects off the white painted concrete floor and gleaming equipment. The cheese room is a model of cleanliness. Before going in, everyone dons white rubber boots, a white hair net, and a fresh white apron.

Inside the cheese room, Jackie runs tests to make sure the milk is antibiotic free, checks temperatures, adds cultures, and removes cheeses from draining baskets. Explaining as she works, her professionalism makes it look easy.

Warren Meadows is her favorite cheese—to make and to eat. This medium-soft cheese is rolled in a blend of rosemary, savory, and thyme and aged for sixty days. The delicate taste of the herbs carries through the flavorful cheese. Aurora and

Vermont Brabander are both harder cheeses, made from cow's milk.

With Jackie, Heather, and Francis doing so much, what do their parents, Linda and Larry, do?

During the three years the family has been getting their flock established and their recipes perfected, there have been other aspects of their sheep business. The Faillaces import sheep for themselves and other Vermont farmers, originally bringing lambs from Belgium and the Netherlands, more recently from New Zealand. Larry holds a doctorate in animal science, is president of the Vermont Sheep Breeders Association, and teaches reproductive physiology. He and Linda oversee all aspects of their family farm. Besides Jackie, Larry is the family's other cheesemaker.

Linda is the family lamb obstetrician and coordinates the business end of the farm. With a partner, she is opening East Warren Schoolhouse Market, a country store located in the 1897 schoolhouse on the Roxbury Mountain Road in Warren. With organic meats and produce from their partner organization, Rootswork, the store will feature cheeses made by Vermont cheesemakers. Starting this week, they'll be the first store to offer Warren Meadows Cheese, made by Three Shepherds of the Mad River Valley.

In the midst of running the farm and store, the house we were renting sold. Back to Anna Whiteside. This time she found us a beautiful rental house, about two miles from the farm, that was built into a hillside and had a sod roof. The interior was spacious with three bedrooms, three bathrooms, a large garage/basement, and two ponds. If you stood on the roof you could overlook the farm, and with binoculars, you could even see the sheep—we felt blessed.

11

THE SILENCE IS BROKEN

In early October we had a phone call: "Linda? This is Dr. Detwiler. ABC's *20/20* filed a request with the USDA for information about BSE a few weeks back, and your information was included. Now I've heard that your local CBS station might have something on as early as tonight."

I was silent for a moment, unsure whether this was good news or bad. Detwiler continued, "They might not even mention your sheep, but we don't know."

"Is there anything we should do?" I asked.

"Just stress the fact that we are all working together, working to find a resolution, looking at this very scientifically. But most of all let them know we are cooperating. Hopefully this whole thing will just blow over."

I paused again, unsure what to say. "Thanks for calling us."

"Just wanted to give you the heads up," she said. I hung up the phone and stood in the kitchen staring out the window at the changing autumn leaves, my mind racing. This was the same woman who almost frantically insisted we stay quiet and sternly reminded us so at every meeting. Now she sounded exceedingly calm, almost blasé. I watched as red and golden leaves fell from the maple tree into the pond and drifted like little vendor boats clustered together at the Thai floating market. What would all this mean? Part of me was thrilled that we could finally go public, tell everyone what was truly going on with the USDA and let Anne Burling and our fellow Rootswork members know about the quarantine. But another part of me was nervous. Detwiler had repeated her warnings about going to the media so many times I was apprehensive of the public's reaction.

My thoughts were interrupted as Larry walked in the door, back from checking on the sheep. I told him about the phone call, and he suggested we contact Anson Tebbitts from the local CBS station.

Anson had reported on the arrival of the sheep, interviewed Freddie about cheesemaking when he first came to Vermont, and did another news piece about our cheesemaking courses. I ran upstairs and found Anson's business card. "Call me on the cell if you need to reach me right away," he once told Larry and me. Within the hour Anson was at our house.

We told Anson all the information we had been forced to keep hidden for the past fifteen months: the USDA's concern that the sheep could have eaten contaminated feed, the fact that we had all the feed records, the fact that the animals were lambs when they were imported and had only been fed grain in the USDA-approved quarantines, the voluntary and then formal quarantines, the test results, all the meetings with Detwiler, Senator Leahy's office, the meeting in Washington, the USDA's refusal to make us an offer, and the resulting standstill.

It was the top story on the local CBS station's 6:00 news. "Hysteria in Vermont over possible mad cow disease," the anchorwoman said. She warned of the seriousness of mad cow disease and how thousands of cows in Britain had succumbed to the disease (viewers throughout Vermont, New York, and Quebec then saw the famous footage of the emaciated cow stumbling in a pen with concrete floors and a metal gate). The reporter talked about CJD but claimed it was "caused by eating infected beef," and then viewers saw photos of our cheese facility with Larry and Jackie making cheese followed by a short clip of Larry saying our sheep were healthy and he would not hesitate to eat the meat. The report then showed more shots of our cheeses juxtaposed with the staggering cow. The newsperson concluded by saying that the USDA was concerned our sheep were contaminated with BSE and should be destroyed, even though there were no signs of illness.

We sat in stunned silence.

The news traveled like wildfire and soon our days were consumed with giving interviews to national and local papers, television stations, and radio stations. Larry and I had given a few interviews over the years, but nothing prepared us for the bombardment once the story was made public. My family stepped in to help. My youngest sister, Becky, a recent international business school graduate, organized our first press conference. My father and another sister, Monica, provided additional scientific expertise. My other sisters lived farther away and offered moral support, and my mom sorted and filed the articles, all while working to keep everyone positive.

One of the first phone calls we received was from the public relations director at Cabot Creamery, the largest cheesemaking company in Vermont. This was not surprising, because we were members of the Vermont Cheese Council, but I was shocked when the spokeswoman attempted to convince us to give up the sheep. Somehow she knew that the beef industry was involved and suggested we take whatever money was offered and be done with it. Although we were never told a dollar amount, she claimed she had heard through the Vermont Department of Agriculture that the USDA would offer $5,000 per animal.

"If you want more money than the USDA is offering," she told me, "you should use arbitration. But don't get lawyers involved. That will slow things down." I thanked her for her suggestions and told her we would not be selling our animals. By the tone in her voice I could tell she was displeased with my response. She wanted us to give up the fight for fear that any additional publicity would hurt the cheese industry.

Bob Paquin was also not happy. "Once you get reporters involved," he said, "you inject a different dynamic." He still hoped that the USDA would agree to "further the science by monitoring the animals."

John Dillon, a reporter working for a local paper, interviewed Detwiler and asked about the "political pressure" she alluded to. Detwiler admitted the issue went beyond the beef industry and included the pharmaceutical industry but gave no further details. She erroneously told John that sheep culled from *both* farms (Mr. Freeman's and ours) had vacuoles, "small holes that could possibly indicate a spongiform disease."

Alfonso Torres told John Dillon, "The high stakes . . . mandate very conservative measures if there is a possibility of the sheep being infected with the BSE agent. . . . However, this is a case in which the welfare of our nation must be placed above any other consideration."

Dr. Tom Pringle, webmaster of mad-cow.org and founder of the Sperling Foundation, wrote the following opinion on his website: "The USDA's action seems like over-kill given the millions of tons of contaminated feed, animal products, medicines, and exposed blood donors already brought into the US over the previous fifteen years of the BSE epidemic. And what about all the sheep imported from 1980–1996? Clearly this is driven by beef export politics."

The politics were about to get worse.

Vermont Commissioner of Health Dr. Jan Carney was angry, really angry. The bumbling secretary of agriculture, Leon Graves, had made a fool out of her. Why hadn't Leon told her about the quarantined flocks of sheep? When a reporter asked her opinion on the matter, she couldn't respond because she didn't even know there was a problem—here in her own state! Carney was appointed secretary of health in 1989 by Vermont Governor Madeline Kunin, and she took great pride in her position. Well, *she* would see to it that the USDA got those sheep.

Meanwhile, the USDA decided the upcoming meeting with Jan Carney was the perfect time to try out their new tactic—divide and conquer. Detwiler arranged for Mr. Freeman and Tom Amidon to meet with Carney, followed by a separate meeting with Larry and me. First thing in the morning on October 27, 1999, Larry and I met with a Burlington lawyer. Our close family friend, Roger Hussey, joined us. Standing around 5'10" and in his early sixties, Roger had an unruly head of white hair and a brilliant mind. We had always confided in Roger, and he was extremely supportive and generous. At one point, Roger took his entire savings and hired a local filmmaker to produce a documentary about our situation. On the day of our meeting with Carney, I was thankful for his calming presence.

An hour later we met the Freemans and Tom Amidon for breakfast and discussed the fact that there was still no word out of Europe, and no cases of BSE in sheep. With high hopes we readied ourselves.

Carney reminded me of a smaller Janet Reno. Standing about six feet with short dark hair and angular features, she spoke in a slow, fairly monotone, but deliberate manner.

Larry took a tape recorder out of his pocket and placed it on the table. Immediately Carney stiffened and the room went quiet. "I will not allow this meeting to be taped," Carney said in a controlled, incensed tone.

"I don't see why not. There is nothing legally stopping us," our lawyer replied.

"If you attempt to tape, we will not meet," Carney insisted.

Leon Graves agreed. "Let's keep this informal," he said.

Bob Paquin, who was also at the meeting, nodded his head. "Yes, let's keep this informal."

Carney spoke up again. "We will only be here about an hour, and I want things to stay collegial and cooperative."

Our lawyer picked up the tape recorder and handed it to Larry who put it back in his suit pocket. There was a collective sigh.

"After talking with Professor Vanopdenbosch I was not reassured," Carney said, anxious to begin the meeting. "He couldn't exclude exposure to rendered products; also, there are the six reports of abnormal brain lesions."

"What are these six reports you are talking about?" I demanded. Over the past few days, Carney and Detwiler had often referred to six questionable test results but until this point had not identified the animals. Detwiler took out a stack of papers and read the animals' ID and accession numbers, which Roger and I wrote down. All the animals were from Skunk Hollow Farm.

Carney was concerned about the flock histories but had not followed Larry's recommendation to contact Dr. Carton. When I told her about the feed records from the originating farms and the Belgian Ministry of Agriculture, Carney requested copies.

Detwiler, who had remained fairly quiet until this point, quickly changed the subject. She had been offered duplicates of the same documents at the meeting in March, but refused. Plus, all the information that was requested from Belgium had been sent to her, and she now had a growing pile in her office.

"The origin flocks have been in the European scrapie program since—" Larry began.

"The animals need to have a five-year certification, and they have only—" Detwiler started.

"Well, the animals could have eaten contaminated feed, and I don't want to take any chances," Carney said haughtily, interrupting both of them. "The cattle in Belgium had records which showed they didn't eat contaminated feed, yet they still got BSE."

"What about the fact that the United States imported meat and bonemeal from Britain right through 1997?" I demanded, staring at Carney.

Carney gasped. "Is this true?" she asked Detwiler.

Detwiler looked down at the table. "Well, we can't give a 100 percent guarantee that rendered product was not imported into the United States," she replied quietly. Detwiler knew about the imported meat and bonemeal because she reported to the Scientific Steering Committee for the BSE risk assessment, and they had uncovered the exact figures of U.S. imports of meat and bonemeal from the UK.

"And what about the fact that BSE has been around since 1985 and now, fourteen years later, there are still no cases of BSE in sheep?" I asked.

"There are a number of experiments going on right now, testing sheep," Carney replied. Research was constantly happening, but what mattered were results, not theories. Before I could respond, Carney switched subjects. "What are your most important goals?" she asked Larry and me.

"Keeping our family and business together," I said.

"And what if I'm right? Then we have a serious situation," Carney replied. "I want the flock to go to the USDA."

"You are talking about irreplaceable sheep," Larry said. "Our sheep are the healthiest and most scrutinized in the world. Kill them, and all sheep are suspect."

"Well, the testing is inconclusive," Carney said, "but the sheep are at risk."

"This 'risk' is entirely hypothetical. If you consider our sheep at risk for eating contaminated feed, you will have to consider all American sheep at risk, since the import figures are evidence that UK meat and bonemeal was imported and probably fed to them," I said. "And what about the fact that the UK sheep that were tested for BSE were not the same breed? Maybe you should speak to Dr. Piet Vellema."

"This is a different diagnosis on your sheep," Carney replied. "The findings are suggestive of a neurodegenerative process."

"They are *not* 'suggestive of a neurodegenerative process,'" I argued. "The changes that were observed on the histopathogies from the Skunk Hollow sheep were nonspecific findings and possibly artifacts from the tissue preparations."

"But these are new diseases we are talking about," Carney replied, again ignoring me. "A human disease called new variant CJD, which is always fatal."

I was getting annoyed with her condescending attitude. "I know about nvCJD," I said, "but what does that have to do with our situation?"

"It's a bad luck of the draw," Carney replied, half smiling.

"Are there any other circumstances, any other options?" our lawyer asked.

"If we are forced to give up the sheep, I would like to see them returned to Belgium," I said.

Larry agreed. "I don't want the sheep killed. But I would like to see more research and testing. Let's come up with an agreed upon protocol, something that will settle this whole thing."

The room was silent as everyone looked at Detwiler and Carney. Neither of them spoke. The last thing the USDA wanted was a compromise. They wanted the sheep dead and to wash their hands of the entire situation.

"So what about the sheep going back to Belgium, what sort of compensation would there be?" our lawyer asked.

"Belgium might even be willing to pay something for them and to use them for more research," Bob Paquin said. He turned to Detwiler and Carney. "What if Glickman declares an emergency? Is the Belgium option viable? Will there be an appraisal of the sheep?"

"We will have to see," said Detwiler.

Despite anything we said, Carney believed the sheep had a variant form of TSE and wanted to declare them a public health risk. No amount of scientific evidence would sway her decision. It was obvious she knew very little about TSEs, but she did not want a dangerous disease coming into Vermont under her watch. Not only did she want the sheep killed, she had decided to go after our cheese. She wanted it off the market.

"Are you selling cheese?" Carney asked.

"Yes," Larry replied.

"How much do you have, and what is it worth?"

"At most about a thousand pounds of cheese, and it sells for eighteen to twenty dollars a pound."

"My recommendations are for you to give the sheep to the USDA and to not sell cheese for two weeks."

"What are the rights of Three Shepherds?" our lawyer asked. "Does Dr. Carney have the right to make an order subject to judicial review?" The meeting felt like it was turning into a free-for-all.

"We will explore the option of buying the cheese," Detwiler said to Carney, ignoring our lawyer's question.

"No, *we* will see about the cheese," Carney replied resolutely.

"Milk and cheese are the FDA's turf," Detwiler said firmly. Carney should have known she was treading on dangerous ground. The USDA controlled plants and animals, while the Food and Drug Administration (FDA) controlled the feeding of humans and animals. There were constant power struggles between the two agencies.

We knew that three weeks before, Leon Graves had sent a letter to Mr. Freeman asking how much cheese his farm produced, how much was on hand, how much was sold, and to whom. But we also knew the letter stated: "Like you, we have been operating under the understanding that there are no concerns about TSEs being transmitted through milk. The World Health Organization has issued a letter to support that conclusion and we understand that it is on that basis USDA allows the importation of cheese from TSE-affected countries."

Now it was our turn to get angry. Never mind the fact that our sheep were healthy, there was no scientific evidence to show TSEs being transmitted through milk or milk products. In fact, the milk from English dairy cows infected with BSE was never recalled, either in the UK, the United States, or elsewhere. Thousands of pounds of cheese from Britain entered the United States every day. Sheep's cheese from Europe was legally sold in shops around the country, and here the Vermont Commissioner of Health was going against every worldwide regulatory agency's recommendations and suggesting we might have to destroy all our cheese. But her suggestion was enough to get the politicians battling.

Governor Howard Dean would not be pushed around by Jan Carney and was much more approachable the next time Mr. Freeman met with him. Dean told Mr. Freeman that if Carney was going to destroy all our cheese, she would have to ban the importation of cheeses from any country that had BSE. This could result in major trade wars. And that was the last thing Dean or anyone else wanted.

That evening Carney was featured in the lead story on the local CBS affiliate, and the segment opened with "Vermont's health commissioner says the sheep should go." According to Carney, "The first thing is that I don't think we can—any of us—can exclude the possibility that this original flock had an exposure to feed contaminated with a BSE agent. The second concern is that there have been at least six sheep who have had autopsies and studies of their brains that have shown what I could call an unexplained nerve disorder."

When asked about the cheese she was now whistling a different tune: "From my perspective and from the CDC's perspective, any risk is theoretical and there is [sic] absolutely no indications for a recall of those products."

Okay, so the risk with the cheese was theoretical, what about the risk of the sheep eating contaminated feed and contracting BSE, wasn't that also theoretical?

The report concluded with statements from Larry expressing his disappointment in Dr. Carney's recommendation and saying we would not sell our sheep without a fight. The following news story was about the New England Dairy Compact getting saved. The compact was a financial mechanism for giving New England dairy cow farmers more money for their milk because of higher production costs in the region. The government controls the price of milk, and New England was finding it difficult to compete against the agribusiness farms of the Midwest. The politics continued.

On November 1, 1999, Governor Dean went public with his support of us and in a radio interview said that various government agencies were at cross purposes, that he did *not* want the sheep killed, that there was *no* public threat, and that it was not clear to him what was behind the scare tactics. It was wonderful to finally get some public political support.

A month later a nationally aired interview by Vermont Public Radio's Susan Hanson shed some more light on the issue for the public. When Susan asked Dr. Joe Gibbs of the NIH why he thought the USDA was targeting our sheep, he replied, verbatim, as he did to us at the meeting earlier that year in Washington, "I wouldn't hesitate to eat a leg of lamb from the Faillace's sheep, but you have to understand that if the United States was perceived to have BSE it would cause the stock market to crash."

Support came from everywhere. A man from northern Vermont wrote a letter to the editor of one of the local newspapers, expressing his disbelief in the situation and how the USDA was handling it:

Recent news bulletins have reported the determination of the U.S. Department of Agriculture (and its satellite agencies, the Vermont Departments of Agriculture and Health) to extirpate two herds of Friesian sheep currently harbored at two Vermont farms. The reason given is that these sheep, having been imported from Belgium, could harbor the agent that causes Mad Cow Disease and represent, therefore, an imminent threat to our lives. I am not an epidemiologist, but to me this sounds far-fetched indeed.

The sheep were imported several years ago under permit from the same department that now wishes to destroy them. Logic dictates that, therefore, either the department was wrong then, and right now, or right then, and wrong now. In either instance, we have proof of the department's incompetence. Prior and subsequent to being imported, the sheep were held in approved quarantine facilities and, for that matter, are practically in a quarantine-like situation now. And in all this time not one of the sheep has developed the disease. What was the purpose of quarantine if the department now suggests that the sheep may be infected?

The department's decision apparently is based on an unlikely set of hypotheses and conjectures: 1) that while in Belgium the sheep *might* have been given feed that contained animal proteins; 2) that, if this were the case, some of these animal proteins *might* have come from cows infected with the disease; and 3) that, therefore, some of the sheep *might* harbor the agent that causes it.

The department's decision process literally begs questioning. For example: Has there been contact with the Belgian authorities to determine if there has ever been a documented case of the kind of transmittal the department now fears? Has there been an investigation to determine whether Friesian sheep in Belgium are normally given feed containing animal proteins? Has the agent that causes Mad Cow Disease been identified, and, if so, is it possible to test animals to determine if they are carriers? Obviously, if the animals are not now carriers, since they now live on Vermont farms they cannot become carriers in the future.

To be sure, the department has offered to reimburse the owners of the sheep. But the reimbursement, in the imaginably most generous set of circumstances, would only return out-of-pocket expense. The future profits to be derived from having been pioneers in the establishment of sheep dairying in Vermont, from the sale of breeding lambs, for example, would be discounted; and, of course, the gratuitous cruelty of putting to death entire herds has not even entered into the department's calculation.

I suggest that there is an entirely different agenda at work here: More and more frequently, Europeans reject American agricultural products because our animals have been fed hormones, or because our crops have been genetically engineered. Further, Europeans subsidize their farmers, and America charges that this is contrary to the GATT agreements. There is an agricultural trade war in the offing: the Friesian sheep may be its first casualties.

Not only were total strangers vocalizing their agreement with our stance, but the Warren selectboard also sent a letter to Dr. Carney:

We, the Board of Selectmen of the Town of Warren, have become aware of some rather disturbing practices on behalf of the U.S. and Vermont Department of Agriculture and the Vermont Department of Health regarding the Faillace sheep farm here in Warren. The Faillaces have extended themselves and their family to restore sheep husbandry to Vermont. They have an imported flock of Belgian Friesian sheep particularly suited to the production of milk and meat in Vermont's climate. They have developed several innovative technologies for the care and maintenance of these animals. Incomprehensibly, this operation has come under great pressure from both the Vermont Commissioner of Health and the USDA to destroy their animals on the suspicion that the sheep may harbor Mad Cow Disease, ostensibly because it is claimed that they *may* have been exposed to feed containing animal protein in their native Belgium. Although the Faillaces have documented evidence to show that this never occurred, and have commissioned scientific studies to show that these sheep have in no way been affected, the governmental agencies concerned have relentlessly pursued the destruction of this family business without scientific evidence to show a justifiable health concern. While we applaud the Department of Health's and USDA's concern and responsibility for public health and the need to take all *reasonable* precautions to keep BSE out of the country, the action in this case smacks of a political expediency that, absent any shred of direct evidence of exposure or infection, should have no place in public policy.

We submit that such practices have no place in Vermont. The Faillace's operation is a model alternative to bovine dairy farming and has great potential to preserve the State's family farm heritage and its working landscape. This family has worked very hard, in good faith to do a good thing for themselves, the town and the land.

We, the Selectmen of the Town of Warren, consider this an important and worthwhile economic enterprise in our small town. In addition, it has far reaching consequences for the future of farming in Vermont. We do not believe that any of the governmental agencies involved have the justification or the authority to destroy the animals or enterprise without concrete physical proof of an imminent danger to public health.

Clearly the USDA wanted the sheep either killed or shipped back to Belgium but lacked any "concrete physical proof" they could use against us to legally force us to comply.

But that didn't stop them.

12

THE THREAT

By our fourth year of lambing we had more of a routine. The ewes were bred in late November or early December so they would lamb on pasture around the time of the first grass in the spring. Lambing outdoors was better for all involved. When the ewes were close to lambing, Larry would check on them more often during the day and every two hours in the night. He never complained. To him this was all part of living his dream.

The Beltex were the last to lamb, and they decided to all do it at once. May 5, 2000, was a beautiful spring day with temperatures in the low seventies. Larry and I had spent the day cleaning out the solar barn and setting up paddocks in the pasture for the sheep. The girls rode the school bus to the four corners that afternoon while Francis stayed after school for soccer practice.

Jackie and Heather raced up the road to see the East Friesian lambs running and spronking in the fields. Only one Beltex lamb had been born so far, but this was soon to change, very soon. As we watched the East Friesians and their lambs in the distance, Jackie noticed Uitgefester looking like she was in labor and ran over to her. "She has some mucus hanging out," Jackie called to Larry and me. Soon there was a nose and some feet.

"Mrs. Friendly's in labor!" Heather yelled from across the paddock. And the next thing we knew so was Olga.

Heather sprinted to the barn to get the supplies while Larry, Jackie, and I each monitored a ewe. Mrs. Friendly was the first to lamb and gave birth to a beautiful large ewe lamb.

Uitgefester and Olga both started squatting at the same time. "Try to move them away from each other," Larry said as he helped Mrs. Friendly's lamb find the teat. But it was too late. Olga had given birth and Uitgefester immediately began to lick the lamb. This stimulated Uitgefester's contractions and soon she gave birth to a ram lamb.

Before we could do anything, Olga had another lamb. Meanwhile, Uitgefester cleaned Olga's first lamb and ignored her own ram lamb.

I picked up Uitgefester's lamb and placed it in front of her while Jackie grabbed Olga's lamb and laid it beside its twin. Uitgefester almost stepped on her own lamb as she shoved past me to get back to Olga's lamb, which she had been licking.

Heather ran back from the barn, her arms full of supplies, and stopped suddenly, a look of amazement on her face as she saw four new lambs. "Whose is whose?" she asked. But before we could answer, Uitgefester squatted and gave birth to another lamb!

Mrs. Friendly was settled with her ewe lamb, Olga was happy to have her second lamb, while Uitgefester claimed Olga's first lamb and her own second lamb. But Uitgefester's ram lamb was abandoned. No matter what we tried—rubbing the ram lamb against its twin to cover both of them in the birthing fluids, holding Uitgefester while trying to get the lamb to nurse, Heather holding the two lambs Uitgefester accepted while Jackie and I offered her the ram lamb—nothing worked. This was one disadvantage of lambing on pasture: no individual pens to force bonding.

Finally, Larry held Uitgefester while Jackie and Heather helped the ram lamb suckle. At least it would receive the precious colostrum—the first milk to come in, which is rich with the mother's antibodies that lambs need to survive. "Maybe if we leave them alone for a while, she will accept him," Larry said. "Let's go home and get cleaned up." We gathered the unused supplies and walked toward the road where our friend and neighbor Doug was pounding a sign into the ground.

"You wanna buy a house?" he called over. We weren't quite sure what he was saying and hurried closer. As we got nearer we saw the sign, For Sale by Owner.

"Are you selling your house?" Larry asked excitedly. We had often talked about how Doug and Alison's house would be ideal but never mentioned anything to them as they were our friends and had just purchased the house a few years ago.

"Yeah, are you interested?" Doug jokingly asked.

"How much are you asking?"

"For the house, apartment, and two acres: $150,000."

Larry and I looked at each other and back at Doug. "We'll take it," we said in unison.

Doug laughed. "Don't you want to take a look at it first?" We had never been inside the house.

"We have to go pick up Francis from soccer practice," Larry told him. "When would be a good time to give us a tour?"

"How about tomorrow morning, say 10:00?"

"Sounds great!"

After six years of searching the entire state, we found our dream home—right beside the farm! Larry and I had no idea how we would pay for it. Up to this point our credit rating was great, but our savings was gone and we lived off the small income from the store and the cheese courses because we could no longer sell breeding stock. But somehow we knew we would make it work. That night as I packed for an upcoming conference on Creutzfeldt-Jakob Disease, Larry talked with his parents and they generously offered to buy the house and have us pay them back. It was perfect.

The next morning we walked through the main house, which was a mid-1800s cape. A spacious wooden deck with a hot tub joined the back of the house with a hundred-year-old barn that had been moved to the property in the 1960s. The first floor of the barn had been converted into a two-car garage with a work area, and upstairs was a one-bedroom apartment built for the former owner's teenagers. Larry and I chose to live in the apartment and let the children have the house to themselves.

The rest of the ewes lambed easily and successfully, but we were still concerned about the ram lamb. Every day Larry would hold Uitgefester while the girls helped the lamb suckle, but Uitgefester refused to allow the lamb near her the rest of the time. One night when Larry and I drove down to check on the sheep, I shone the flashlight in the field and noticed a small bundle, huddled alone, far away from the rest of the flock. It was the ram lamb.

"We'd better take him home," Larry said.

"But then he will end up being a bottle lamb," I sighed. Larry and I were proud of the fact that in four years of shepherding all the ewes had been able to raise their lambs. I thought of Benedict and Uitgefester. Well, maybe not their own lambs, but they raised all the lambs. A bottle lamb would require multiple feedings day and night. And where would we keep him?

"It's too cold tonight, and he might not make it," Larry said. So we took the ram lamb home and put him in the sunroom with our rabbit and guinea pig. Jackie named the lamb "Moe."

I flew to Miami the following weekend for the first conference of the Creutzfeldt-Jakob Disease Foundation. My goal was to talk with Dr. Mary Jo Schmerr about a blood test for TSEs she was developing. If her test was successful, all our sheep could be tested easily without being killed, and the test kit would cost less than twenty dollars once it was approved. A protocol could be established and the animals tested once a year for several years. After the specified time period passed, the quarantine would be lifted, and we would have our lives back.

I took a taxi to the waterfront hotel, and the woman at the front desk happily informed me she had upgraded my room to a king-sized bed with a water view. The room was gorgeous. I wished Larry was there to share it with me, but we could only afford for one of us to attend the conference, and that was putting a squeeze on the budget as it was. But if it saved our sheep, the money would easily be worth it. After I changed out of my traveling clothes and put on a suit, I wandered outside the hotel enjoying the warm sunshine.

That evening was registration and a reception. I met a man in his thirties who was interested in learning more of the science behind CJD because his young wife had it, and another man in his fifties whose wife had died of the illness a few years earlier. Over the course of the weekend I was shocked to meet so many people who had lost a loved one to CJD. A surveillance system for Creutzfeldt-Jakob Disease had been established by the Centers for Disease Control (CDC) and Dr. Pierluigi Gambetti at the Institute of Pathology at Case Western Reserve University in Cleveland, Ohio. The goal of the surveillance was to determine if CJD had increased in the United States. But researchers faced an uphill battle because U.S. hospitals were not required to report CJD, and, from the stories I heard, doctors were discouraged from listing it on the death certificate. Mandatory reporting was one of the goals of the organizers of the conference.

Shortly before lunch on the second day of the conference, I finally had an opportunity to speak with Dr. Mary Jo Schmerr.

"You don't want me to test your sheep. I'm having problems with false positives and false negatives, and my procedure is not yet validated," she told me in a hushed voice, as her dark eyes searched the meeting room to see if anyone would overhear our conversation. Of medium height with short brown hair, Dr. Schmerr was a former nun and maintained a quiet, gentle manner. She was very friendly and soft-spoken; her mannerisms reminded me of my Jackie. This was someone working for the USDA whom I instinctively felt was trustworthy.

"I'm still working on the test," she continued. "It's not ready." Disappointed, I explained the story of the USDA trying to seize the sheep. Dr. Schmerr listened intently, nodded sympathetically, and asked if we had Dr. Katherine O'Rourke test our sheep using her "eyelid" method. Sheep have a third eyelid, the "nictitating membrane," which helps keep dust and particles from getting into their eyes. Scrapie is believed to begin in the intestinal system, make its way into the lymphoreticular system, up the spinal cord, and finally into the central nervous system via the brain stem. Once the disease reaches the brain stem, the infected animal will begin exhibiting symptoms. Because the nictitating membranes are part of the lymphoreticular system, Dr. O'Rourke developed a test that could detect PrPres, the abnormal prion and indicator of a TSE, at earlier stages in the progression of the disease and long before an animal would show clinical signs. More important, a snippet of eyelid could be tested on live sheep, meaning all of our sheep could be tested and would only suffer a little discomfort.

Although the USDA used the eyelid test on the mastitic ewe from Freeman's farm, each time Larry and I asked Detwiler to use Dr. O' Rourke's test on our sheep, her response was always the same: "The test is not ready yet; it's not validated."

Later that afternoon, John Stauber, coauthor of *Mad Cow USA*, and Dr. Michael Hanson from the Consumers Union spoke of the inadequacies of the FDA and the USDA in dealing with TSEs. Around 650 biological products that came from cattle were used in everything from supplements and cosmetics, to insulin and pharmaceuticals. A country that has BSE is no longer able to source their own cattle population for these products The nutritional supplement industry in the United States was worth $14 billion. On top of this, the beef industry was worth more than $40 billion and gained enormous political power with their large political contributions. The beef industry contributed more to politicians' coffers than the tobacco industry. As I listened to Michael and John, I decided they might have some advice for our situation.

Unsure whether they would be supportive, I gathered my courage and approached the stage at the end of their talk. Most of the audience was leaving for the day, but a few of the conference attendees remained in the room talking among themselves. I climbed the steps and offered my hand, "Dr. Hanson? Mr. Stauber? I'm Linda Faillace. I own the sheep in Vermont that the USDA is trying to seize."

Both men shook my hand. "So you're the one giving the USDA hell," John said with a smile. I felt my shoulders relax.

"That's right," I replied. "We are trying to work with the USDA to resolve this whole issue, but we seem to be up against a wall."

Just then a larger man in his late forties rushed onto the stage and pushed past me. "How can you say that the USDA and FDA are not being proactive enough?" he demanded of John and Michael.

"Excuse me?" Michael said.

"If you want to talk about risks to BSE in this country, you should be talking about those Vermont sheep," the man said irately.

"I am—" I started to say.

"Those sheep are more dangerous than all the risks you mentioned," he interrupted.

"Excuse me—" I again attempted.

"*We* were talking," the man turned angrily, looking down his nose at me.

"Well, I happen to be the owner of one of the Vermont sheep flocks you are talking about," I replied, now getting defensive.

That caught his attention.

"And you are?" Michael asked the man.

"Lawrence Schonberger of CDC," the man replied. There were no warm handshakes here as he turned his full focus on me. "How can you sleep at night knowing you are putting the entire American livestock industry at risk with your sheep?" he asked, glaring at me.

"I sleep quite well because there is nothing wrong with our sheep," I answered, unable to keep the snide tone from my voice.

"How can you say that?!" he asked in exasperation.

"Because there is nothing wrong with our sheep," I repeated, unflinchingly.

Schonberger turned to John and Michael. "Don't you think they should kill those sheep?" he asked.

"I haven't seen anything yet that would convince me," Michael replied.

"Every single test the USDA has run on every single sheep has been negative. The USDA is so intent on scrutinizing our sheep that they have run more tests on our sheep so far this year than on cattle in the United States," I said, standing tall at 5'2".

"Just who do you think you are?" Schonberger asked, sneering at me. "You think you know more than the scientists here? You think you understand this?"

"I can understand the science of BSE, yes," I said. I definitely had the passion and interest and had been studying TSEs for almost ten years, I thought to myself.

Schonberger changed his tactics when he saw he wasn't getting very far. "Well, what would it take for you to give up those sheep?" he asked, leaning in toward me, close enough for me to see the sweat on his brow.

"We won't give up perfectly healthy animals, because there is nothing wrong with them," I answered, refusing to back away.

"How about if you were offered more money?" he asked.

"It's not about money," I said.

"What if your sheep tested positive?" he demanded.

I hesitated. I had never considered this possibility. "Well . . . I guess we would have to surrender them then," I said quietly.

"Great!" Schonberger exclaimed and rushed off stage.

Michael turned and put a hand on my shoulder. "Are you okay?" He could see I was shaken.

"Thanks, I'm fine," I said, trying not to show how upset I felt.

"That was strange," Michael said. "Did you hear what he said? I'd watch out for him."

13

SEIZE AND DESTROY ORDER

Larry, Jackie, and Freddie were scheduled to teach a cheesemaking course in Maine in early July, followed by two more cheese courses at our farm. In addition, our cheeses would be featured on the *Martha Stewart* show, and we were hosting the monthly Vermont Cheese Council (VCC) meeting at our farm. Life in the cheese world was exciting.

Before Jackie and I left to get Freddie from the Boston airport, I received a phone call from Detwiler saying she wanted to meet on July 17th at the Vermont Department of Agriculture. When I asked about the purpose of the meeting, Detwiler was vague, so I phoned Bill Smith.

"What is this next meeting for?" I asked.

Bill sounded fairly upbeat, "It's time to wrap this whole thing up. It's gone on for too long." I interpreted this as a positive response and breathed a sigh of relief—maybe we could finally get back to our normal life. Cheerfully I recounted the conversation to Larry and the children. On the way to Maine, Jackie was noticeably happier, even talkative. It was as if a burden had been lifted from her tiny shoulders.

When we shared the good news with Freddie, he put his arm around Jackie and said, "Now we can focus on the important things in life, like cheesemaking!" Jackie and Freddie had a strong bond, and Freddie was very proud of his young protégée who, by the age of eleven, was making cheese on her own. Freddie pulled out a large package and handed it to her.

"This is for you," he said. "Now you are a real cheesemaker."

Jackie quickly unwrapped the cardboard packaging, and her face lit up. It was a cheese "harp," a stainless-steel, multibladed, curd-cutting knife.

"Thank you, Freddie!" Jackie said, beaming as she hugged him. This was the only child I knew who got excited over stainless-steel utensils.

You could be sure Jackie would enjoy a birthday or Christmas if a large, commercial-style, stainless-steel spoon was among her gifts.

After a delicious breakfast the next morning at a B&B in Harmony, Maine, we drove to the Maine Organic Farmers and Gardeners Association fairgrounds. The cheesemaking course was full with ten students. I handled the introductions, helped set up, and then drove home to Vermont to prepare for the VCC meeting. The next day Francis and I mowed the lawns and weeded the flower gardens while Heather ran the store. Francis and Heather continued milking the sheep twice a day.

On July 12, 2000, our cheese was presented on *Martha Stewart*. The piece focused on the children and the incredible job they did with the animals and cheesemaking. That evening I received a phone call from Dr. Todd Johnson from the Vermont Department of Agriculture. When Sam Hutchins retired as the Vermont state veterinarian, former New York state veterinarian Todd Johnson took over his position. Todd informed me our meeting at the Vermont Department of Ag would be the following day, not on the 17th. I winced and explained to Todd that Larry was in Maine and would not arrive home until later that night and that we were hosting the VCC the next day. Todd was insistent. I finally relented and agreed Larry and I would be at the Ag department at 4:30 P.M.

I was annoyed. I hated the way we were expected to drop everything at the last minute, just to fit Detwiler's schedule, or whoever was responsible for this latest change. "At least it should be over soon," I told myself.

That night I expected Larry, Freddie, and Jackie to be home around 10:00 P.M. Eleven o'clock went by, then midnight. This was unlike Larry. If he was going to be late, he would call. Finally at 12:30 A.M. the phone rang. The drive back from Maine was taking longer than expected, and there were few phones along the sparsely populated route. It was raining, and Larry had already passed two moose in the road and was driving slowly in case there were any more. I told him to hurry home safely.

Shortly after 2:00 A.M., the truck pulled into the driveway. Jackie and Freddie said quick good nights and went straight to bed. I was a bundle of nerves. "Why didn't you call me earlier?" I demanded of Larry.

"I never thought it would take this long," he said, moving toward me for a hug.

"You knew I was waiting and must have known I would be worried," I insisted, stiffening my shoulders when he put his arm around me.

"I'm sorry. I'm home now," Larry said with a slight smile, knowing that his smile was usually the fastest way to ease the tension. I didn't smile back. I was extremely tired and wound up about the upcoming USDA meeting.

"USDA says we have to meet tomorrow at 4:30 at the ag department," I told him.

"What do you mean tomorrow?! I thought the meeting was supposed to be on the 17th?" Larry asked, his grip tightening on my shoulder.

"I know, I know," I said, exhaustion taking over. "I tried to change it, but he was firm."

"Who did you talk to?"

"Todd Johnson."

"And what was his tone like?"

"It's hard to say. He always sounds like he doesn't want to be on the phone. I don't know, I guess he seemed a little more serious than usual." That's what was really niggling at me. Something didn't feel right.

After milking the sheep the next morning, Heather and Francis opened the store while Jackie helped Larry and me prepare for the VCC meeting. Vermont is blessed to have people who are passionate about food. We have incredible bakers, cheesemakers, winemakers, vegetable growers, livestock farmers, and plenty of market opportunities for all. The Vermont Cheese Council was formed to bring cheesemakers together and help promote their wonderful products. I was pleasantly surprised at the camaraderie among the members. The council included everyone from the larger producers like Cabot Creamery, Grafton, Vermont Butter and Cheese, and Shelburne Farms, to small operations like the two women who milked five cows in the winter and made a small amount of cheese. All the cheese council members worked together: sharing recipes, buying things in bulk, and helping promote each other's cheeses. Because the market base was so extensive, there was little competition among the cheesemakers.

At the Schoolhouse Market, along with our cheeses, we sold cheeses from Lazy Lady Farm, Green Mountain Blue, Willow Hill, Vermont Butter and Cheese, and Grafton. All the cheesemakers from those farms came to the meeting that day, including the VCC president, Alison Hooper of Vermont Butter and Cheese.

When everyone arrived at the farm we gave a quick tour of the store and the cheese facility. Yestermorrow, a local architecture school, was in the process of dismantling the old porch and building a new one on the front of the schoolhouse, so we began our meeting outside at the backyard picnic tables, and when the work crew was too noisy, moved upstairs in the schoolhouse.

There were more items on the agenda than usual, and the meeting was lasting longer than I had anticipated. Nervously I watched the clock. At 3:30 I leaned over to Larry and whispered that we had to leave. We stood and explained that we had a meeting at the ag department and apologized for the early departure. Everyone smiled and thanked us for hosting the meeting.

Larry and I were both quiet as we drove to Montpelier. We did not know what to expect, but we were not feeling optimistic. Roger Hussey had offered to attend the meeting with us, and we met him in the lobby of the ag department. Waiting in the meeting room were Detwiler, Leon Graves, Todd Johnson, and Michael Duane from the Vermont Attorney General's office.

The moment we sat down Detwiler shoved papers toward Larry. "Four of Mr. Freeman's sheep tested positive," she said.

"What?!" Larry exclaimed.

"Four animals from Skunk Hollow tested positive," Detwiler repeated. "I have Dr. Rubenstein, who performed the tests, on the speakerphone here."

"Richard, are you there?" she asked.

"Yes, I'm here," a man's voice answered.

"We have the same players as this morning," Detwiler told him. (They had met with Mr. Freeman and Tom Amidon earlier that day.)

"Since when were we *players*?" I thought. "Does Detwiler think this is all a game?"

With a smug look on her face, Detwiler turned her attention back to Larry and me. "Dr. Rubenstein ran some Western blots and confirmed that four animals had PrPres." She continued by explaining the Western blot procedure, which Larry and I were already quite familiar with: Brain samples were taken and treated with an enzyme, proteinase-K, which digests all the normal prion protein (PrPsen). If any proteins (PrPres) resist digestion, they appear as a band on the blot and the animal is considered positive.

"Did you rerun the samples?" I asked Rubenstein.

"Yes, I did," Rubenstein replied.

"Where were the tests performed?" Larry asked.

"Only in my lab," said Rubenstein.

"Is your procedure validated?"

"Yes."

"Do you still have more material left?" I asked.

When Rubenstein hesitated and cleared his throat, Detwiler interjected, "We still have tissue. APHIS has the frozen brains."

Larry and I sat in silence. I noticed Larry studying Rubenstein's report closely. So did Detwiler. "What did you mean that Rubenstein's test 'confirmed' that four animals were positive?" Larry asked Detwiler.

Detwiler rummaged through her files and pulled out more sheets of paper. "Dr. Mary Jo Schmerr tested the animals and four of seven samples were positive."

I looked with disbelief at the papers in front of Larry. "But I spoke with Dr. Schmerr in May, and she said her test was not ready," I told Detwiler.

Detwiler did not expect this response and quickly changed the subject. "You have until noon tomorrow to surrender your animals," she said.

"What?!" Larry and I gasped.

"This is out of my hands now. It's at the [ag] secretary's level," Detwiler responded, a triumphant look on her face. "Will you voluntarily surrender your animals for fair market value?"

Larry and I looked at each other. I shook my head "No."

"We are not ready to make that decision yet," Larry told her.

"Well, the alternative is that the USDA will make that decision for you and file a 'Declaration of Extraordinary Emergency.' This will send it straight up."

Detwiler continued, "You realize that we can't sit on this story. We are going to declare this a potential TSE to the media."

I felt myself glaring at her. All this talk about keeping a low profile, making sure to stay out of the press, and now she would run to the media claiming the sheep had a TSE.

"This is impossible," Larry said, still looking over Rubenstein's papers. "No East Friesian in the world has ever even had scrapie."

"Now, we are going to have to consider the environmental effects," Detwiler said, again changing the subject. "USDA will bring in a panel to give advice, to decide the disinfection method. We are considering taking the top six inches of topsoil off areas where the sheep were. Anything which came in contact with the sheep during lambing will also be destroyed."

"You have got to be kidding!" I cried out. This would include our dog and llama. "What about the fact that none of our sheep tested positive?" I demanded.

"We have to assume that your animals were infected, since one of the four which tested positive was an original import," Detwiler said. "You received lambs which were born at Skunk Hollow, did you not?"

I nodded, and my heart sank.

Leon Graves spoke up from the end of the table, "Have you sold any meat?"

"No," Larry replied.

"I want to inventory the sheep," Todd Johnson added.

"All the animals are tagged except for one lamb," Larry told him with resignation.

Detwiler stood to signal the end of the meeting. "You realize that this is a matter of days now, not months," she told us.

"There's something wrong with this test," Larry said, looking over Rubenstein's papers.

My dad agreed. He and my mom had rushed over to our house as soon as they heard the news. With a PhD in biochemistry, former executive officer for two laboratories, and founder of a diagnostic testing laboratory, my father was well qualified to review test data.

"If you look at Dr. Schmerr's test and compare the samples to Rubenstein's, the animal that tested 'most' positive on her test tested negative on his, and an animal that tested negative on her test was positive on his," Larry continued. "And her positive control was negative!"

"And you notice that Rubenstein doesn't draw any conclusions from the test results, he just presents his 'findings,'" my dad pointed out.

Freddie, my parents, Larry, and I sat on the deck at our house strategizing. Moe was cuddled on my lap like a puppy. Every so often he would look up at me as I talked and then go back to resting his head on my leg. As I held his warm wooly body, I felt shivers rush up my spine. The USDA wanted to kill him, as early as tomorrow. I needed to keep it together. Now was the time to focus, to fight, not to break down.

"Can I see Rubenstein's and Dr. Schmerr's papers?" I asked. Larry handed them to me. I read, "As per your request . . . Vermont sheep . . ." on both results.

"What about the fact that Rubenstein and Dr. Schmerr knew where the samples were from? They should have been unlabelled. Wouldn't this mean they would have had a confirmatory bias?" I asked. Both Larry and Dad nodded their heads.

"This definitely seems like a setup," Larry said.

"Do you have any information on these particular animals?" my father asked.

"Not yet," I replied. "Mr. Freeman said the Whittens are trying to match their records to the USDA numbering system."

"What about the fact that Detwiler is taking this to the press?" he asked.

I was sure the USDA planned the timing. The worst time to get something into the news is on Friday, particularly Friday afternoon, because most reporters were slowing down for a well-deserved weekend break. And fewer people watched the news on Friday evening or read the papers on Saturday. If the meeting had taken place on Monday the 17th, the press would have been fresh and raring to go.

"Maybe we should beat her to it," I said.

Larry mulled it over for a moment and then said, "That's a good idea. Why don't we arrange a press conference for first thing tomorrow morning and share all the test results. Let the public make up their own mind instead of being misled by pseudo-scientists from the USDA."

"Are you still awake Hon?" Larry asked softly.

"Yeah, I can't sleep," I said, rolling over to face him. All I could think of was Moe, Mrs. Friendly, Upsala, and all the others and how we were responsible for their lives. Then I thought of our children. Tears welled in my eyes. I wished I could take away all their pain. They were too young to have to go through this. Surely this would affect their outlook on life. I always tried to focus on the positive and teach the children that anything was possible, but this was totally out of control, out of our hands. I felt like a mother hen, wanting to gather them under my wings, hide them from what was happening.

"I think the USDA should re-run the tests," Larry said, interrupting my depressing train of thought. "There are too many things that raise questions. If the USDA re-runs the tests and they come back positive, then we will have to give up the sheep."

"I would agree to that," I said. "And I think the children would too."

By six the next morning, the entire family and Freddie were up and dressed. Larry and I explained the press conference, what we would be asking of the USDA, and encouraged the children to feel free to talk about their feelings to the journalists.

"Don't do anything to get arrested, Mommy," Heather said to me with tears filling her dark eyes.

I hugged her and laughed. "What makes you think I would do something that would get me arrested?"

"You really love Moe and will fight anyone who tries to take him from you."

She had a point.

"That's true, Honey, but I love you guys more, and I won't do anything to hurt you or the rest of the family," I said.

Francis, however, was ready to give a piece of his mind to anyone who would listen. He was furious over Detwiler's threat to take Pélé. "Try to stay calm," I urged him.

"How could they do something like this? Where are our rights?" he demanded. "This is not what I learned the United States is all about. We are supposed to be the best country in the world. We are supposed to have all these rights. That's just bullshit."

"Francis! Don't use that language!" I yelled.

"Well, it is," he said, crossing his arms. "I think we should move to Canada."

"Moving to Canada is not going to solve this," I said. But the thought had crossed my mind.

"We should be leaving for the store now," Jackie whispered. "I want to be ready for the cheese course students."

"Oh my word! I forgot about the cheese course!" I exclaimed. How could I? Freddie was there. I enjoyed organizing the cheese course, and yet, with all that was happening, it had totally slipped my mind. Freddie stood there smiling. "Jackie and I can handle everything," he reassured us.

Roger and Bruce Fowler met us at the schoolhouse and offered to help run the store and make lunch for the cheese course so we could focus on the USDA. The two of them had already been busy calling all our friends, letting them know what was happening. Soon the parking lot was full with reporters, cheese course students, and local supporters holding placards and handing out "Save Our Sheep" bumper stickers. We decided to have the press conference behind the cheese facility so Freddie and Jackie could talk to the students in front of the schoolhouse.

By 9:30, Larry, Francis, and I stood on a small knoll with more than a dozen reporters surrounding us. Most of the reporters had interviewed us many times already and were quite familiar with our story.

"The USDA claims that four sheep from Skunk Hollow Farm tested positive for a TSE," Larry began. "The USDA used a lab that is not on their list of laboratories approved for running TSE tests, and there were several obvious flaws with the testing. We have until noon today to surrender our animals. We are asking the USDA to meet with us to discuss these results and to re-run the tests."

The questions lasted for forty-five minutes. When the conference was over, Larry pulled me aside. "Do you see the man by the Jerusalem artichokes?" he asked.

A small gravel road wound from the parking lot of the schoolhouse to the cheese facility with a row of Jerusalem artichokes lining one side. Standing a few yards away from the group of reporters was a man in a dark coat, clearly out of place.

"What about him?"

"Do you know who he is?" Larry asked.

"No. I have no clue."

"I have a bad feeling. He didn't take any notes, just stood in the back, talking on his cell phone." I thought it was funny that in the midst of the chaos, Larry would notice one particular person.

The man then looked up and saw Larry and me staring at him. He nodded his head toward us, turned, and walked to the parking lot.

"Do you want me to find out who he—?"

"Linda? I'm Graham Johnson from WPTZ, could we ask you a few questions over here?" A young man approached me with his microphone outstretched.

I looked at Larry, wondering what I should do.

"Go ahead, talk to Graham. I'll check in on Freddie and Jackie."

When the interview was over, I saw Larry talking to more reporters and decided to see how Heather was doing. She was in the store handing out cheese samples while two of her and Jackie's friends helped run the register. The Yestermorrow students were cutting boards and finishing the floor of the new porch. And the phone was ringing off the hook.

When Heather finished selling cheese, she rushed over to me and handed me a sheet of paper. "You need to call some of these people," she said. "They all want to talk to you, *now!*"

I looked at the list: the *New York Times*, National Public Radio (NPR), CNN, and more. Heather had the list organized by time of call, name of caller, media organization they represented, and a blank column to check off when their call was returned.

"You've done a great job with this, Heather!" I exclaimed.

"Thanks. But will you start calling them?" she urged.

I ran upstairs to check on Freddie lecturing the cheese students. It was quiet and felt safe from all the turmoil happening below.

Back downstairs, I called Mr. Freeman. He said Senator Leahy had phoned from Washington to try to delay the USDA, but it was too late. Mr. Freeman stressed we should all work together, and I wholeheartedly agreed.

Next, I called Detwiler.

"Dr. Detwiler? We have reviewed the test results and would like to have the animals retested."

"I'm sorry," she said. "I'm just a peon in this whole thing. You will have to call Secretary Glickman." That was not good. It was already 10:30.

"Do you have his number?" I asked. She did. Immediately I called his office.

"Secretary Glickman's office," the receptionist answered.

"Hi, this is Linda Faillace, and I need to speak to the secretary," I said.

There was a pause. "I'm sorry. The secretary is in a meeting right now. Would you like to leave a message?"

I could feel my palms start to sweat. "No, we are the owners of the sheep in Vermont, and it is very important that I speak to him as soon as possible. He is probably talking about our sheep in the meeting."

"Sheep in Vermont? I'm not sure what you are referring to," she replied.

"Look, we are under threat of losing our sheep if I don't talk to the secretary before 12:00."

"I'm sorry, ma'am. But the secretary will be in meetings till then."

"Great," I thought. "Now they are trying to tie our hands. Give us an answer by 12:00, but we won't have anyone you can talk to, so it won't matter anyway."

"I need to talk to someone," I insisted.

"Would you like to speak with Dr. May?" she asked.

"Sure," I said. I had never heard of a Dr. May, but at this point I would talk to anyone.

"I'll transfer you."

"Hello, this is Butch May."

"Dr. May?" I said, "This is Linda Faillace."

There was a pause. When he didn't say anything, I continued.

"I am one of the owners of the sheep in Vermont."

Still no response.

"The sheep that the USDA is trying to accuse of having mad cow disease . . ."

"Really?! Sheep with mad cow disease? Your *sheep* have mad cow disease?"

"No! Our sheep do *not* have mad cow disease. There are no sheep in the world that have mad cow disease, but the USDA is trying to claim ours might."

"Wow!"

"If I don't get ahold of Secretary Glickman before noon, the USDA is going to seize and kill our sheep."

"Wish I could help you. Did you try calling his office?"

"Yes, that is who transferred me to you."

"They transferred you to *me*?" he asked, almost incredulously. "Did you tell them who you are?"

"Yes, I did, but they said the secretary is unavailable."

• • •

A few minutes later, Larry walked into the store, followed by three reporters with their notepads, microphones, and cameramen. "How did it go? Did you call Detwiler?"

I told him how I was getting the runaround. "Maybe I should call Dr. Torres," I said. "At least he was sympathetic."

Larry rolled his eyes. We had had this debate before in Washington. Larry felt Dr. Torres was putting on a front, while I thought he was sincere. "Whatever works," he said, grimacing.

I called Dr. Torres and told him about Rubenstein's tests, the flaws, our proposal, and how I was unable to reach Secretary Glickman. When I started talking about the science behind TSE testing, Torres cut me off and told me I would have to talk to Detwiler. This annoyed me. Torres was supposed to be Detwiler's boss, but he was deferring to her.

"Do you mean to tell me that Detwiler makes all the decisions?" I asked. "She said she is a 'peon in this whole thing,'" I continued. "So, if she can't help us, you can't help, and Secretary Glickman is in meetings until noon, what are we supposed to do?"

Torres had no suggestions.

I could feel the tears starting as I hung up the phone. I felt trapped. And then Francis walked in carrying Moe. "I thought you would like to have him with you," he said.

"Thank you," I said as I hugged both of them and took Moe into my arms. Once I leaned down and smelled his wool, the tears rushed out. I carried Moe into the garden shed and sat with him on a milk crate. He started bleating. He wanted to scamper down the ramp that ran from the garden shed into the backyard and eat some grass. I smiled as I set him down and watched his chubby little body hurry down the wooden slope.

Twelve o'clock came and went. No word from Glickman's office. Some reporters were camped out at the picnic tables, others watched the cheesemaking, while a few in the store documented the phone calls. I called Mr. Freeman to see if he had heard anything from the USDA, but he hadn't. The phone rang again, and Heather answered it.

"Mom, there's a woman from NPR on the phone and she wants to talk to you or Daddy," Heather said, her hand covering the mouthpiece.

Larry listened to NPR every day. "Let Daddy do that one," I said with a slight smile. I had often teased him about being addicted to NPR.

Then the other phone line in the garden shed rang, "Linda, what should we do?!" It was Robin Aldrich, the owner of the third flock of sheep. Robin and her husband Barry were the only other farmers besides us and the Freemans to hold out against surrendering their sheep to the USDA.

"I don't know, Robin," I told her. "We are all going to stick together. I tried getting hold of Glickman today, but his secretary claims he's in meetings."

"Do you have a lawyer?" she asked.

"No," I said. "We have been working with Mr. Freeman and his lawyer, Tom Amidon." The Burlington lawyer we had hired for the meeting at Carney's office was more interested in having us surrender the sheep and reach a financial settlement than fighting the USDA for what we felt was right, so the relationship had ended quickly.

"Will Tom Amidon add us on if we go to court?"

"I don't know. Hopefully we won't have to go to court. I'll keep you posted."

There was a long pause. Robin cleared her throat. "Can you get me Roger Ives's phone number so we can have semen collected from our rams?" she asked with tremendous sadness in her voice.

Even though it was now late in the afternoon I tried Glickman's office again. The response was still the same, "He's in meetings." On my fifth call a different receptionist answered. When I asked to speak to Secretary Glickman she said he was unavailable, "But . . . you could speak with Dr. Craig Reed. He's the deputy administrator for APHIS."

"Fine, could I have his number please?" I said. She gave me his number.

Dr. Reed was available.

I explained who I was and that I was calling to have the USDA re-run the tests. He quietly listened and then said, "What tests do you want them to re-run?"

I sensed that Dr. Reed was not very knowledgeable about TSEs and decided to question him. "What tests do you use to determine if cattle have BSE?" I asked.

"Um, you will have to talk to Dr. Detwiler. She is our TSE expert. Would you like her number?"

At this point I felt ready to scream. "You use a histopathology and an IHC," I said angrily. "And no, I don't want her number!"

I continued, "If you are going to test cattle in the United States for BSE, you use those two tests. If both of those tests are positive, the animal is considered positive." I took a deep breath, "The USDA is claiming that sheep from here in Vermont tested positive, but it was by a Western blot. Not a test the USDA uses, and it was run in an unapproved laboratory."

"A Western what, ma'am?"

"Western blot." I could tell I wasn't getting anywhere.

"I'm sorry I can't help you, ma'am."

So was I. "Yeah, thanks anyway," I said, trying to keep the sarcasm out of my voice.

Just as I was about to hang up, Dr. Reed spoke up, "Ms. Faillace?"

"Yes?"

"Secretary Glickman has already signed the seize and destroy order."

14

HELL, NO, DON'T KILL MOE!

Detwiler wasted no time getting interviewed and spreading misinformation. From an article written by reporter John Dillon for the Montpelier paper, the *Times Argus*:

> "In the East Friesian breed there's never been a diagnosed case of a TSE," Linda Faillace said. ". . . So this doctor [Rubenstein] has found the first case of TSE in this breed in the world . . . if his test is correct."
>
> But Detwiler said she was mistaken, saying that at least one case of Friesian sheep being diagnosed with TSE had been documented in England.
>
> And Detwiler said the new version of the chemical test was scientifically accepted.
>
> "It's a . . . test that's been validated worldwide," she said. "The procedure has been validated and published in peer review journals, and the test results (on Freeman's sheep) have been confirmed twice."
>
> Detwiler also disputed that the doctor who ran the test knew where the samples came from when he tested them. . . .

Detwiler must have been feeling pretty bold to claim the samples were anonymous. Rubenstein's report, addressed to Detwiler, stated: "As per your request, we processed these *Vermont sheep* brain samples for Western blot analysis" (my emphasis).

Detwiler also skirted the validation issue by referring to Western blot tests in general, saying it was a "test which has been validated worldwide." If this Western blot was so widely accepted, why was it not used by the USDA for their diagnostic testing?

Some Western blot procedures had been validated. But each Western blot used a different antibody to bind with the proteins, and each procedure with a different antibody had to be validated independently. Rubenstein used his "rabbit anti-C57BL mouse PrP antibody," which had *not* been validated.

I had met Jonathan Leake of the *London Times* at the Miami CJD conference and yet the only information source he used for his article was Detwiler, and his article was full of erroneous statements and dangerous exaggeration.

"Vets have found the first evidence that sheep may have become infected with BSE," he wrote. "Three flocks in America were this weekend being forcibly taken from their owners to high security laboratories for slaughter and testing after tests showed clear signs of the disease."

The sheep were not seized, and there were no signs of any disease.

He continued, "The flocks of East Friesian milking sheep were imported from Europe to Vermont within the last four years. Checks have shown that their parent flocks had eaten British-bought feed that was likely to contain material from BSE-infected cows."

We held all the feed records, documented and signed by the flock owners, the feed mills, and the Belgian Ministry of Agriculture. And none of the feed came from Britain!

"Linda Detwiler said . . . 'Four sheep were confirmed positive on July 10 for a transmissible spongiform encephalopathy (TSE)—the class of diseases that includes BSE. Specific tests for scrapie, the TSE normally found in sheep, have proved negative so there is a distinct possibility they have BSE instead.'"

This defied logic: If an animal tests negative for a disease by all validated, normal testing procedures, how could they be positive for an even more serious disease, especially when the tests for scrapie and BSE were the same?

And this was the last thing that Britain needed—a threat of BSE in sheep. If the disease were to show up anywhere, you would assume it would be in the British sheep population, because Britain had by far the most cases of BSE in cows and their sheep would have been exposed to the highest percentage of potentially BSE-contaminated feed. Farmers were still reeling from the effects of the massive slaughter programs, the public fear of eating meat, and record-low meat prices.

Jonathan Leake also wrote, "Four scientific tests were used on the

Vermont sheep. The first looked at the brains of culled animals and found clear signs of lesions seen in sheep with BSE-like diseases. When tests designed specifically for scrapie proved negative, the samples were sent to other scientists to see if they could find prions—the protein particles that are thought to cause and transmit TSEs."

The USDA never found BSE-like lesions. Leake didn't seem to understand that prions didn't cause and transmit TSEs; PrPres, or "abnormal" prions, were believed by some scientists to be the culprits. And who were the scientists sent on the wild goose chase to see if they could find abnormal prions in samples that had already tested negative?

"A USDA scientist who carried out one set of tests said several of the animals had proved positive—and her results had been independently confirmed by another scientist using a different technique."

Here was a science writer who had not verified his facts. Dr. Schmerr's test was still in its experimental stage (the positive control tested negative, not positive), and Rubenstein's test was *not* validated. His laboratory was not even on the USDA's list of approved laboratories to test for TSEs.

"'The sheep are negative for scrapie but they definitely have some kind of prion disease. It could be a new form of scrapie—or BSE,' she [Detwiler] said. 'The evidence in favour of BSE is strengthened by the animals' European parent flocks having been certified completely free of scrapie.'"

Again this made no sense logically or scientifically. The animals tested negative and their parents were certified free of scrapie, so therefore they must have an even more threatening disease? According to Leake, Detwiler said it would "take a year" for further tests to confirm whether the sheep had BSE. So which was it, the "several years" as Detwiler had told the Associated Press and all the major American media sources, or "a year" as she had told the European sources?

And to make sure total panic set in, Leake concluded his article by claiming "Americans may be at risk from vCJD since milk from the sheep has been sold for human consumption as has a limited amount of meat."

The USDA, CDC, FDA, and NIH had all stated that there was no evidence of TSE transmission through milk and milk products. Yet according to Detwiler and Leake, the American population was now at risk of contracting a horrific, fatal disease from our sheep—a disease that our sheep didn't have, which did not exist in sheep, and which was transmitted in a manner that no scientist had ever been able to demonstrate as possible.

I wasn't the only one outraged by the article. Dr. Tom Pringle posted extensive comments on his website (mad-cow.org) to Leake's piece. He argued against the USDA's "possibly BSE" conclusion and also pointed out the politics behind the USDA's actions:

> Now the USDA never gave two hoots about game farms importing and exporting mad elk or hunters eating animals known to be infected with CWD. Americans have long had the USDA's blessing to go on eating sheep with scrapie . . . so, why a multi-million dollar monitoring program with black helicopters sweeping in to seize three small flocks of sheep?
>
> Let's put it this way: the US doesn't export billions of dollars of meat, cosmetics, and pharmaceuticals made from elk or sheep.
>
> USDA is a little schizophrenic about the sheep industry. On the one hand, a $100 million emergency subsidy, plus huge new bailouts of the wool and mohair lobby. On the other hand, the USDA is destroying the sheep industry with disastrous worldwide publicity saying mad sheep reached the US in 1996 and you may have already eaten one (or had the cheese or milk).
>
> Where is the thirty-five-member SWAT team that was lined up to minimize the damage from the first announced case of BSE in US cattle? It all boils down to sheep and cattle are both to receive ag subsidies but when there is a conflict, cattle rule. Indeed, the USDA has equated the welfare of the cattle industry with that of the nation: "The high stakes involved mandate very conservative measures if there is a possibility of the sheep being infected with the BSE agent," Alfonso Torres, deputy USDA administrator, told the Faillaces [the Vermont flock owners] in a letter this spring.
>
> "This is a case in which the welfare of our nation must be placed above any other consideration."
>
> Conveniently, "further testing, which will take several years, is required to determine which type of TSE has infected these

sheep." USDA is left with a debatable TSE diagnosis, and the BSE call looks like total nonsense.

It doesn't matter: they [USDA] issue themselves an order to forcibly seize and incinerate the sheep and have a fair shot at winning a court battle. If they're wrong, so what, the owners received fair market value [see below]. Beef-based cosmetics and pharmaceutical exports can resume from a country riddled with TSEs but one that has taken draconian and highly visible steps to avoid even far-fetched BSE. Mission accomplished.

"The owners will receive fair market value." Had the owners obtained permission to distribute the breedstock, their value would be astronomical since the import window has closed and the sheep are unique. Without this permission, they are just assessed as sheep that will produce a small amount of milk relative to a cow and aren't worth much at slaughter—what are they worth now as USDA-certified BSE mutton?

Shortly after the situation hit the international headlines, Pélé went missing. Francis was close to tears as I tried to console him. "Maybe he ran to Lucky Dog," I suggested. Lucky Dog was the kennel Pélé would stay at when we traveled. It was about half a mile from our house, and whenever Pélé wanted to go out for a jaunt he would visit Duke, Laure, and all his buddies at Lucky Dog.

"I called them and Laure said she hadn't seen him," Francis said with a look of dread in his eyes.

"He'll show up soon," I tried to say reassuringly, wrapping my arms around him. I thought about the USDA's threats. Pélé was sitting beside us as we told this to the media at the media conference. Did he somehow understand? Or did the man with the cell phone have something to do with it?

Dark thunderclouds were moving up the valley from the southeast, and a gusting wind rattled the windows of the schoolhouse. Francis pulled away and hurried out to check on the sheep. We would have to delay the evening's milking until the storm passed. The tension of the day's events and the fact that I hadn't slept much over the past forty-eight hours finally caught up with me. I called Moe in from the garden behind the school. He pranced up the ramp and ran over for a hug. I held him close and closed my eyes, listening to the thunder rumble in the distance.

Throughout the stormy weekend, Larry and I spent most of our time on the phone. Mr. Freeman was in contact with Governor Dean and Senator Leahy. Larry called Dr. Mary Jo Schmerr and asked about her results. Dr. Schmerr pointed out the fact that her positive control was negative, and the findings were therefore invalid.

"These results would never hold up in court," she assured Larry. We had no intention of going to court and found her comment odd, but just brushed it off.

Then Larry called Madeline Kunin, former Vermont governor and ambassador to Switzerland. I spoke with a woman from Savannah, Georgia, who told me she and her sister in Arlington, Texas, were in contact with Governor Howard Dean and Governor George Bush. Our local representative, Kinny Connell, spoke impassioned pleas to Governor Dean, but it fell on deaf ears. People around the country offered their support: Whom could they write letters to? Could they protest? Send money? Supporters were willing to do anything to help.

Roger Hussey was in his element. He became the self-appointed head of SOS (Save Our Sheep), set up an office in the store, and used the chalkboards to list news items and daily lists of things to do. He established a phone tree for supporters, in case the USDA showed up at the farm, and created press releases.

Larry and I designated Heather "media coordinator," and she excelled at it. She carried a clipboard to keep track of all the calls and interviews and created a phone tree of reporters to call in case the USDA arrived. Everyone was busy. There was no time to stop and think about what was happening, no time to feel, no time to cry.

Carton called. We were the headline story on the Belgian news. Europeans speculated why the USDA would accuse our sheep of having a disease that did not exist. Some thought it was due to the fact that the EU wouldn't accept U.S. hormone-laden beef; others claimed the U.S. was jealous that the EU had better sheep.

Similar conversations were taking place here in the United States. The theories about why this was happening were rampant. Did we upset someone in the sheep world? Was Detwiler trying to gain attention and perhaps a promotion? At meetings and in interviews she had often spoken of her father's pigs being seized and killed by the government for hog cholera when she was young. Was this a bizarre form of retaliation? Or was it a smokescreen for the USDA's woefully inadequate BSE cattle-testing program? By publicly seizing and destroying our sheep, the focus was taken off the powerful cattlemen's association and put on the tiny sheep industry. They could destroy a flock without huge market consequences for the beef and pharmaceutical industries, which practically control the USDA and FDA. And yet at the same time the USDA could send a message to the world that they were being tough on BSE. One thing on which everyone agreed was that our animals were being victimized by power groups in this country and sacrificed by bureaucrats.

All I knew was I wanted to save our sheep. I asked Carton if he would check into the sheep going back to Belgium. In November of 1999 the USDA had sent a letter to Belgium asking if they would allow the return of the sheep. Larry and I had told Carton to let his ministry know we were attempting to find a satisfactory resolution with the USDA, so no action was taken at that time. Carton suggested we now speak to Mr. Adriansens of the Belgium embassy in Washington. We prayed Belgium would welcome them back home.

On Saturday afternoon a member of the press, not the USDA, faxed us a copy of the "Order to Dispose of Sheep." A few reporters had arrived early to the store and now stood anxiously, waiting for our reaction as Larry and I silently read the order together.

> Sheep and Germ Plasm Requiring Disposal: The entire flock of sheep, identified by scrapie flock ID VT003, which are currently under quarantine at RR1, Box 122-3, Warren, Vermont 05674 and germ plasm from these sheep.
>
> Under section 2(c) of the Act of July 2, 1962, 21 U.S.C. Section 134a (1962) you are hereby ordered as the owner of the above described sheep, which are affected with or exposed to an atypical transmissible spongiform encephalopathy (TSE) of foreign origin, to allow United States Department of Agriculture (USDA) officials to seize and dispose of these animals and their germ plasm on or before July 21, 2000.

You will be compensated the fair market value for these sheep and their germ plasm in accordance with 21 U.S.C. Section 134a (1962). USDA will obtain two appraisals of your flock. If you wish, you may also obtain an appraisal prior to July 19, 2000. These appraisals will be used to determine the fair market value of these sheep and their germ plasm.

The disposal of said sheep and their germ plasm is necessary because of the risk these animals and their germ plasm pose to United States livestock.

No liability shall attach to the USDA or any inspector or representative of that Department with respect to the transportation and/or disposal of the above described sheep and their germ plasm.

Done at Washington, D.C., this 14th day of June.

Craig A. Reed, Administrator, Animal and Plant Heath Inspection Service

"What is an 'atypical TSE of foreign origin'?" I asked Larry.

"I have no idea," he said, still staring at the order. "I bet this is a term they came up with to protect themselves legally. If the USDA claimed the sheep had scrapie they would have to follow their scrapie regulations."

If the USDA followed their scrapie rules, only the infected animals and their direct offspring would be killed. The flock would remain under surveillance for a number of years and as long as no additional animals tested positive or exhibited clinical symptoms, the flock would then be cleared. But because none of our animals or Mr. Freeman's had tested positive for scrapie, the USDA could do nothing other than break their own rules.

Larry continued, "And if the USDA claimed the sheep had BSE, they would have to follow the BSE Emergency Response Plan."

The USDA had worked long and hard to develop an elaborate protocol to be followed if and when BSE was found in the United States. The animal would be tested by histopathology and by immunohistochemistry (IHC). If both tests were positive, samples would be hand-delivered to Central Veterinary Services Laboratory in Weybridge, England, where a confirmatory test would be performed.

"That's what they should do," I exclaimed. "Take samples to Weybridge. That would settle this for once and for all."

"You know there's no way the USDA would ever do that. They would lose face."

He was right.

"Why the 'foreign origin' label?" I asked, but I was already forming my own opinion.

"They are taking the blame off the United States and putting it back on Europe."

"And I notice that the dates don't correspond," I said, pointing out the fact that the text of the letter said it was "Done at Washington, D.C., this 14th day of June," while the date stamped at the top of the document said "July 14, 2000." Either the USDA had this planned for more than a month, or they were poor proofreaders. I gave them the benefit of the doubt on that one. But when I saw a copy of Mr. Freeman's seize and destroy order, it was dated "July 18, 2000," four days after ours. Wasn't it his sheep they were most concerned about?

And once again I saw that I had been fed a line. Reed had signed the order, and yet he told me Glickman did.

July 21, 2000—we had five days to save our sheep.

I thought the weekend was busy, but once Monday rolled around life took on an even more hectic pace. Reporters from major national and international networks arrived at the store, and the parking lot was soon crowded with television trucks and their large satellite dishes anchored on top. The second cheese course was finished, but Freddie stayed with us for the next course. Yestermorrow students worked on the schoolhouse porch and watched the goings on with increasing interest. When they had signed up for a class in design and construction they never imagined a major news event would unfold around them.

Moe spent every day as close to me as possible. Because the parking lot was full night and day and I couldn't let him roam free outside the store, the classic old wrought-iron-fenced cemetery beside the schoolhouse was the perfect spot for him to graze while I worked the store or gave interviews. When Moe was very young he could squeeze through two bars of the iron fence that had a slightly larger gap than the rest, but as he got bigger he was surprised to see "the bars had shrunk."

Heather arranged for Larry to interview with *CBS Evening News*.

The Yestermorrow students heard about the upcoming interview and hung signs on the posts of the cemetery fence; signs that read SAVE OUR SHEEP and HELL, NO, DON'T KILL MOE. When the CBS reporter, Jim Axelrod, and his camera crew arrived, the students chanted, "Hell, no, don't kill Moe," while Moe watched from behind the bars, wishing his pudgy little body would allow him to join in all the commotion.

Jim's report:

> The farm's in Vermont, not Britain. The animals are sheep, not cows. But the government's position is familiar: it wants to destroy them.
>
> "It makes us feel obviously outraged, upset, let down, like you really can't depend on the government at all," says Dr. Larry Faillace. He owns 120 of the sheep . . . condemned not because they have mad cow disease, but because tests show they could.
>
> "I realize it's a really emotional thing to the owners," says Dr. Linda Detwiler, a veterinarian with the U.S. Department of Agriculture. "While we're really sympathetic to the owners, we've got to look at the big picture."
>
> The big picture was painted five years ago in England, where cows started turning up sick with a neurological disease. More than fifty human deaths were traced to meat from the cows. The worry is that these sheep ate feed made from sick cows. Dr. Faillace isn't buying it. "Killing a flock of perfectly healthy and valuable sheep is not the way of protecting the American public, and they know it," he says.
>
> Faillace imported the sheep from Belgium for their extraordinary milk capacity. His tests show his sheep are clean. "Slaughtering the sheep is the government's payback for European Union restrictions on American cattle imports," he says. "This is a form of retaliation against the EU. And it's not going to be some foreign disease that's going to kill our sheep. It's going to be politics."
>
> The government wants to slaughter by Friday and is promising fair market value for the sheep.
>
> "We're charged with keeping livestock healthy," says Dr. Detwiler. "So to take that risk to introduce it into livestock

and subsequently the public . . . I think in this case . . . this action is very warranted."

But for this Vermont farm family, it's not about money.

Once again, rural Americans feel strong-armed by the federal government. It seems there's nothing Larry Faillace can do to save his flock.

"Let's take a walk outside," Larry said, taking my hand and moving toward the door in the store as the CBS truck pulled out of the driveway.

"Why?" I asked, my voice fraught with tension.

"Because you could use the break," Larry said gently.

"But . . . I can't. I'm supposed to call back the reporter for New England Cable Network," I insisted.

"It can wait," Larry said firmly. I sighed and followed him out to the parking lot. He was right, I should take a break. Here it was almost four in the afternoon, and I had forgotten to eat lunch.

As we walked in the bright sunshine with a slight breeze whispering through the maples, I noticed how hunched my shoulders were and straightened up, taking a deep breath.

Just then a dark sedan followed by a green SUV with Connecticut plates pulled into the parking lot. The car stopped beside Larry and me, and the tinted windows rolled down. Two men were in the car. "Are you Larry Faillace?" the driver asked Larry.

"Who wants to know?" Larry replied.

Surprised, I looked at Larry. I didn't expect the sharp edge to his voice.

"We're U.S. marshals, and we are here to serve you an order," the man replied.

"What is your name?" Larry demanded.

The man was silent.

"What about your partner? What's his name?"

A pause and then the driver spoke again, ignoring Larry's questions, "The man behind us has the papers." And with that he drove the car forward to the store.

Meanwhile Roger walked outside, saw the conversation taking place, and approached us.

"Everything okay?" he asked.

"The Feds are here," Larry said pointing, "and they don't want to give out their names."

"I hear you, Bro!" Roger winked to us as he approached the two men who remained seated in their car. "I'd like to see your badges," he said authoritatively.

By now the green SUV had pulled up beside us. Larry and I looked at the driver and then at each other. Larry squeezed my hand so hard it hurt. It was the man from the press conference; the one Larry had seen talking on his cell phone by the Jerusalem artichokes. He handed Larry a large brown envelope.

"I assume this is the seize and destroy order," Larry said, taking the envelope.

"Yes."

"And who are *you*?" Larry asked.

"I'm James Finn. I work for the USDA Office of Investigative and Enforcement Services."

"Do you realize how wrong it is what the USDA is doing?" Larry asked.

"Look, I'm just doing my job," Finn replied, obviously wanting to give us the papers and get out as soon as he could.

"You would be wise to research what is going on here," Larry continued. "I would not want to work for an agency involved in abusing science and creating a false public hysteria for political gain."

"Well said," I thought admiringly.

"I'm just doing my job," Finn repeated. "You won't see me around here again."

Larry and I looked at each other. What a strange thing to say.

The federal agents were anxious to get away from Roger's barrage of questions and the reporters now snapping photos. Bruce had watched from the porch and walked over as the vehicles pulled out of the driveway. "Who was that?" he asked, pointing to the green SUV.

"Finn," I said, "James Finn. Works for the USDA. Some office of investigative something or other."

"That guy was in my driveway this weekend."

Bruce now had our full attention. "What do you mean in *your* driveway?" Larry asked.

"He was parked in my driveway," Bruce repeated, "for quite a while, off and on throughout the weekend." Bruce lived about a mile and a half from the schoolhouse, on a fairly obscure dirt road.

I laughed. "These guys are real swift. The driveway they choose to 'hide out' in is my business partner's!"

Roger joined us and smiled. "Well, no names and no badge numbers," he said. "But they will think twice about wanting to come back here."

On Monday evening after milking the sheep, we held an SOS meeting upstairs in the schoolhouse. More than sixty supporters and reporters attended. Everyone came with suggestions, comments, and fears. The biggest concern was the USDA showing up unannounced and stealing the sheep.

As the sun set, I alternated between feeling feisty and energized and feeling overwhelmed. But the negative thoughts were stronger, and it was hard to keep the tears from flowing. The children presented brave faces to the media, but tempers were short in the hot humid air. They were tired of all the interviews, of the tension surrounding us. We all wanted our peaceful life back.

The next day Jan Carney issued a warning, on behalf of the Vermont Department of Health, cautioning consumers against eating cheese from "Three Shepherds of the Mad River Valley," despite the fact that there was no scientific evidence to support her decision. She never bothered to distinguish between our cow's milk cheeses and sheep's milk cheeses. Jackie burst into tears as a reporter read her the warning. Everything she had worked so hard for—the years of learning cheese-making, teaching with her father and Freddie, and the resulting wonderful publicity—would be destroyed by an overzealous bureaucrat.

Seeing her sister in tears was too much for Heather. She had tried to maintain her distance, not get caught up in the emotions, but this pushed things too far. No one was allowed to hurt her sister. For five minutes Heather paced the store, oblivious of the reporters and their cameras. She had to do something. But what? Finally she came over to me and with a firm voice said, "I want to call the White House."

Later that evening I found Francis sitting quietly on his bed, staring out his window, boxes surrounding him. We were trying to move in the midst of everything.

I sat down beside him. "Are you okay?" I asked.

He nodded but didn't say anything as he twisted Pélé's bandana in his hands.

"He'll be all right. He's a smart dog. Maybe he's just lying low," I said. "Or maybe he found a pretty female to visit," I joked.

Francis held his head even lower.

As I leaned over to hug him the phone rang. "Maybe somebody found him, Francis. You should answer that."

Francis didn't move.

After the third ring, I ran to the bedroom and picked up the phone. A man's voice on the other end brusquely said, "We have Pélé, and he's fine," then quickly hung up.

15

FUGITIVES

Mommy, don't get arrested." Heather repeated what was now her mantra as the family sat around the breakfast table. Her call to the White House the day before had not yielded any results, and she was obviously feeling the same sense of desperation I was. The woman who had answered the phone at the White House assured Heather that her message would get passed on but did not know if and when Heather would hear something.

"Honey, I promised you I won't do anything to get myself arrested," I said again, trying to calm her fears. "And I will stick to my promise." My family always came first.

Meanwhile, Francis and Larry discussed how to handle the USDA appraisers who would arrive shortly. When Bill Smith called to say the USDA had hired two appraisers to assess our sheep, I reminded him of the appraisal Larry and I had done in January when we flew over the head of the East Friesian studbook (Mr. Pissierssens) and the head of the Beltex studbook (Dr. Bernard Carton). They spent a week evaluating the sheep and their potential in the American market and wrote a formal appraisal.

"I know," Bill agreed, "but the USDA wants their own appraisal."

"Who will be doing the appraising?" I asked.

"Dr. Dave Thomas and Axel Meister."

Axel Meister?! Axel was a Canadian dairy shepherd and husband of Chris Bushbeck. Chris had taught Larry a nonsurgical artificial insemination technique for sheep a few years before, and we often met up with them at various sheep conferences. Just two months earlier Axel had called to say his barn had burned down and he was in need of money. Would we be interested in buying any East Friesian lambs from him? He had twenty to thirty East Friesian ewe lambs that survived the fire, and he would sell them for only $1,200 Canadian each. We offered our condolences and explained that we

The author's happy family in the winter of 1998. Clockwise from top left: Larry, Linda, Francis, Heather, and Jackie Faillace.

Visiting with Basque shepherds in Spain.

The first importation of European sheep passes the USDA quarantine.

The Faillaces' first farm, in the Mad River Valley, Vermont.

Francis and Heather, the two older siblings, move the sheep to fresh pastures.

The Faillaces' youngest daughter, Jackie, Freddie the guard llama, and the Beltex ewes in winter.

Heather holds Benedict, the first Beltex lamb born in the United States.

All the lambs loved Martha, an East Friesian ewe. Photo by Amalia Elena Veralli

Francis shepherds the Beltex ewes and lambs.

Heather and Jackie milk the East Friesians.

Jackie creates artisanal cheese.

Linda and Moe.

East Friesian lambs dance in the pasture. Photo by Amalia Elena Veralli

Francis carries Moe, a bottle-fed Beltex lamb, to the USDA livestock trailer as police and federal agents swarm the farm.

Heather, Jackie, and Francis hold lambs for the last time.

Linda argues with a federal agent while her sister, Becky, stands beside her.
Sandy Macys Photo

The flock boards the USDA trailer that will take them to quarantine. Sandy Macys Photo

One of the Faillaces' rams, Jacques, is led to the USDA trailer while a policeman looks on.

Protestors block the road in front of the USDA trailer. Sandy Macys Photo

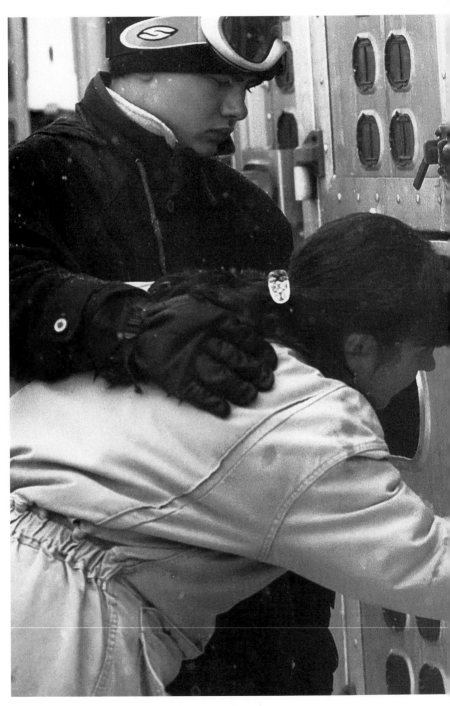

Linda and Francis say good-bye to Martha. Sandy Macys Photo

We will never forget. Photo by Amalia Elena Veralli

Francis, Heather, and Jackie manage the Schoolhouse Market store—and the media— while their parents are in court.

Linda and Larry give testimony about BSE to the New York State Legislature task force on Food, Farm, and Nutrition Policy.

The organic gardens behind the Schoolhouse Market are tended by Rootswork, a group dedicated to promoting community-centered agriculture.

Linda and Larry inside the Schoolhouse Market, where they sell fresh, local farm products.

were under quarantine and unable to buy any animals but said we would spread the word around to some people we knew who might be interested. How could Axel agree to help the USDA? Was it just for the money?

Now Dave Thomas didn't surprise me. Dave was a faculty member at the University of Wisconsin, and his USDA-funded work involved the breeding and evaluating of East Friesian crossbreds. He remained aloof and detached whenever we saw him at different sheep events, and I sensed we were not some of his favorite people.

Over breakfast we decided Francis would show the appraisers the sheep, Heather would coordinate with the media, and Jackie would run the store. The schoolhouse parking lot was filled with reporters, and the family worked quickly to prepare the store for the day. Moe waited for me in the cemetery and nuzzled me contentedly when I knelt down and scooped him up in my arms. Francis joined Moe and me as the USDA appraisers arrived. Out of a large government-issued SUV stepped Wayne Zeilenga, Yves Berger, and Axel Meister. In the second vehicle were several U.S. marshals. I walked through the parking lot holding Moe, staring in disbelief. Dave Thomas was supposed to come, not Yves. Yves was Dave Thomas's technician and someone we respected. Axel looked at the ground, trying to avoid eye contact. "Imagine the money the USDA spent to fly Axel from Canada and Yves from Wisconsin," I thought. "The USDA must have paid really well. There was no amount of money someone could pay me to turn against fellow shepherds."

Wayne walked toward me and shook my hand. He smiled at Moe as cameramen from news agencies around the world jostled to get the best shot.

"Larry and I have a meeting this morning with our lawyer. Francis will show you around the farm and the different groups of sheep," I told him. Francis formally shook Wayne's hand.

Axel, Yves, and Ed Jackson (from the Vermont Department of Agriculture) were dressed in gray disposable coveralls with plastic boots. Wayne and his assistant, Darryl Kuehne, put on their traditional cloth coveralls. No one donned gloves.

Tom Amidon had called and urged us to get a lawyer, because it looked like the only way to stop the USDA from seizing the sheep was to go to court. "Plus your case is stronger than ours because none of your animals tested positive," he said.

We asked Ted Joslin, the lawyer who helped us incorporate our business, to represent us in our battle against the USDA. Ted recommended another lawyer from his firm whom he felt would be even more competent. His name was Davis Buckley, and Larry and I would meet him for the first time that day.

Standing around six feet with brown hair and deep hazel eyes, Davis Buckley resembled movie star Ben Affleck. He greeted us warmly and took us to the conference room where he listened intently as we recounted our story. Larry and I explained how we wanted to stop the USDA from seizing the sheep and force them to perform additional testing. Davis grasped an enormous amount of detailed and complex science in a very short time, and the next few hours were spent writing and rewriting affidavits. Finally Davis suggested we break for lunch. Larry and I walked to Sarducci's, a wonderful Italian restaurant overlooking the Winooski River in Montpelier, and ordered appetizers.

As we ate our mussels and calamari we talked about Davis, our impressions, and the affidavits. Mr. Freeman was filing a separate lawsuit, but we would continue to work together. Our goal was to have the four animals retested. Detwiler said there was plenty of tissue left and using the European rapid tests, Prionics', or even the American company's (Biorad) test, everyone would have an answer in less than twenty-four hours. The animals were under quarantine and therefore were not a risk to the American livestock industry. Everything seemed fair and straightforward.

"But what if the marshals serve us papers?" asked Larry.

"What papers?" I asked, dipping my bread in olive oil. We already had the seize and destroy orders.

"The official papers notifying us of the actual date of the seizure. Once the USDA has the appraisal finished, there is nothing to stop them from taking the sheep."

I dropped the piece of bread. What little appetite I had was now replaced with a knot in my stomach. "You mean to tell me they could seize the animals before we even get to court?"

Larry and I looked at each other, our minds racing. We asked for the bill and rushed back to Davis's office.

• • •

Our fears were confirmed. "As soon as the USDA serves you the papers, there's nothing prohibiting them from taking the sheep," Davis told us. "The good news is Federal District Court Judge Garvin Murtha has agreed to a conference call tomorrow morning. The bad news is you have to make sure you don't get the papers before then."

Larry and I were speechless. All I could think about was the children having to shoulder all the responsibility back at the farm. "Let me call the store and see how the children are doing," I said. Davis suggested I use the speakerphone.

When I dialed the number, Heather answered. "Where are you? I've been waiting for you to come home! When are you coming home?" she cried in a rapid frantic voice. "They said Kiki has mastitis! It's not true, she's my best milker! They don't know anything about my sheep."

"Slow down, slow down, Sweetie," I said, trying to calm her. "What are you talking about?"

"The people with the USDA said Kiki has mastitis and that's why her udder was so big. It's not true, she's my best milker!" Heather sounded close to tears.

"It's okay, Honey. Don't worry about it," Larry said.

"But it's not right!" Heather exclaimed.

"I'll take care of it," Larry reassured her.

There was a loud sniffle on the other end, and then in a quiet voice Heather said, "Well you know what I did?"

"What did you do?" I asked gently.

"I put 'Save Our Sheep' bumper stickers on each of the USDA's cars while they were in the back fields looking at the rams with Francis," she said with a satisfied tone.

"Heather!" Larry and I exclaimed together. We looked at Davis, who was silent and smiling.

"When are you coming home?" Heather asked again. "The marshals keep hanging out at the store and asking me every ten minutes when you will be back."

Larry, Davis, and I looked at each other. There were signs of our phones being tapped: a delay in the dial tone when we picked up, unusual clicking noises, and the occasional sound like someone hanging up.

"We will be home in a while," I told Heather.

"What do you mean 'a while'?" she demanded. "I thought you would be back by now. How long can it take to meet with the lawyer?"

"It's taking longer than we expected, Honey. I'll call you in a bit. Is Roger there?"

"Yeah, he's talking with a reporter right now."

"How's Jackie?"

"She's out in the cheese facility. I think she's hiding." Jackie found it soothing and calming to focus on her cheesemaking. Plus she prohibited reporters from entering her space unless they put on a hair net, aprons, and boots. Jackie was very particular about her cheesemaking area, and some reporters she refused permission to enter.

"And Francis?"

"Francis is out in the field with Kortney."

"Okay, we will call you later," I said. "We love you."

"I love you, too."

As Davis left to pick up his son from preschool, Larry and I walked slowly to the truck, trying to figure out what to do. Our conference call was scheduled for 9:30 the following morning. How could we get home without being seen? Our black Ford F150 with "Save Our Sheep" and "Got Vermont Cheese?" bumper stickers would be easily recognized. As we drove down State Street in Montpelier we passed Commissioner Leon Graves leaving the ag building. This was the last person we wanted to see. He looked directly at us. My heart stopped. "He saw us!" I told Larry. "We have to get out of here!"

We decided to rent a car from Enterprise. We had often worked with the company, leasing cars for Freddie and other visitors, and the occasional rental for ourselves. John, the man who arranged all our previous rentals, was working that day, and he greeted us warmly. "Well, hello! Where are you going this time, off for another cheesemaking course?"

"No, we just need a car for a day or so," I said.

"Where are you headed? Back to New Hampshire to help some sheep get pregnant?" he joked, remembering that Larry taught artificial insemination to sheep and goat farmers.

"Darn," I thought. "Of all the people to rent us a car, we got the guy with the great memory." I worried the marshals would contact the rental companies and try to track us down.

"No, we are just taking a break together," I told him, hating the fact that I felt like I had to be secretive.

"Oh, okay," he said, slightly dejected.

A few moments later he perked up and asked "A romantic break?"

Larry and I smiled but said nothing. Romance was way down on our list at this time.

"So, how does it feel to be the big media stars?" John asked, changing the subject. "I see you every night now on every station. I even saw you on the national news."

I wondered what the news reports would be saying that evening. Would the reporters talk about the fact that we hadn't come home?

Larry and I tried to figure out a way to get home to our children and our farm that night but finally concluded it was too risky in case the marshals were poised there, papers in hand. We drove north on Route 89 and pulled off at the next exit to call George Schenk and explain our predicament. George's daughter Hanna was Jackie's best friend, and George agreed to take the girls to her house for the night. Francis and Kortney were spending their nights upstairs at the schoolhouse. Our friend Bill Carnright had loaned us a service station bell, and Francis spread it across the driveway in order to alert him and Kortney should the USDA trucks enter the parking lot in the night. George said she would have her husband, also named George, check in on the boys. I thanked her profusely.

"So where do you want to go?" Larry asked.

"I don't know. Bec and Joe are away on vacation, we should probably stay out of the valley, and I definitely don't want to go as far as Canada. What I want to do is go home." And with that the tears started. I was angry. Why were we forced to be fugitives in our own state? Forced to flee from the U.S. Marshals? Forced to hide in order to be heard in court?

Larry tried to calm me down. "It's only for the night," he said. "So where do you want to go?" We tried to think of places where no one would look for us and an hour later were sitting outside one of our sheep friend's home in lawn chairs, sipping a glass of wine, telling her and her husband our disturbing story as the sun set over Lake Champlain. They were good listeners and offered support but also felt some fear. They didn't want anything to happen to their animals. Maybe it would best for us to keep our distance.

Darkness was approaching so Larry and I said good night and made our way back to Burlington. We figured we would stay at one of the larger hotels but were dismayed to find out that most hotels were booked. There were quite a few conferences that week in Burlington, and it appeared that every room was occupied.

"Maybe we will have to stay at the Swiss Host Motel," Larry said with a grin, trying to find humor in our surreal situation. The Swiss Host Motel was a tiny 1960s-style motel painted fluorescent green with bright purple trim. It was located in South Burlington among other strip architecture. A very strange and unbecoming place.

For the past seven years Larry would joke every time we went to Burlington that "someday I will take you to the Swiss Host Motel." As we tried hotel after hotel that night I grew depressed. I wanted to watch the 11:00 news to find out what happened on the farm.

At 11:20 P.M. we checked into the Swiss Host Motel.

"The marshals were up at your house when I went to get clothes for the girls," George told me. It was after 11:30 P.M. and I apologized for calling so late, but George just shrugged it off and went back to talking about the marshals. "They were parked at the end of the driveway and watched as I pulled in. It was creepy."

Unbelievable. The marshals staked out our house.

"How are the girls?" I asked.

"They're fine. Pretty tired though. It was a long day for them. As soon as they had dinner, they fell asleep on the couch trying to watch a movie with Hanna and Willis."

"And Francis?"

"He's loving this. What sixteen-year-old boy wouldn't?! He and Kortney are at the schoolhouse. And Freddie is staying with Bruce."

Freddie! I had totally forgotten about Freddie. The poor man was left to fend for himself. "Thank heavens for Bruce and our friends," I thought.

"So what's up with the court?" George asked. I explained about our conference call and what our hopes were. "I'll be here to take care of the girls if you need more time," she said. We were blessed. Blessed with incredible friends, and I thanked her.

Our room at the Swiss Host Motel was actually very clean and comfortable, but we were still unable to sleep. Around 5:30 in the morning, Larry turned on the television. Interviews with the children were on every local station. We lay in bed watching as Francis spoke with reporters about the USDA appraisers, Jackie made cheese, and, last but not least, reports of bumper stickers "mysteriously" appearing on USDA vehicles.

16

"THE FARMERS HAVE NO RIGHTS"

Judge Garvin Murtha agreed for all parties to meet the next after-noon. As Larry and I searched for a parking spot, I saw the media already gathered outside the Burlington federal courthouse waiting for our arrival. It felt surreal, like something I had seen on television, not something I ever expected to experience firsthand. I looked at Larry in the driver's seat and leaned over to hold his hand. He was deep in thought and had a furrow on his brow. I asked if he was okay, and he nodded but didn't say anything. Davis was waiting for us by the entrance. He answered the reporters' questions so well you would have thought we had worked together for years.

Around 1:30 the court clerk announced the case and introduced the lawyers: Davis was our lawyer, Mr. Freeman hired Tom Amidon and Tom Higgins, and representing the government were Melissa Ranaldo, Joseph Perella, and Paul Van de Graaf. The obligatory pleas-antries were exchanged, then the judge immediately reprimanded the USDA attorneys. The USDA was late submitting their filings, and the court (and the lawyers) had not received anything until shortly before noon.

"So obviously I have not had a full opportunity to review them, and I assume that the plaintiffs have not either, correct or not?" Judge Murtha said.

The lawyers agreed.

"And, secondly," Judge Murtha continued, "the purpose of today's hearing, if that's what it is, is to establish what, if anything, the court should do in regard to the requests by the plaintiffs. Now it seems to me that since the government has indicated that it is not going to take any action in regard to the sheep prior to this court's decision, is that—"

Melissa Ranaldo quickly interrupted, "That's correct, Your Honor. At this point in time the government would request that the court rule on the merits as it is before the court. We have asked the court to issue an order to compel compliance with the government's—the USDA's—order, and we would request that that be issued."

"All right. Then—"

But before Judge Murtha could finish, Tom Higgins stood. "Your Honor—if I could for a moment, Your Honor, just comment on the state of the record with respect to the government's representations. There is an order outstanding of the USDA for the seizure and disposal of the sheep, and absent an order of this court in some respect, whether it be a consent order that we enter into with the government, I am reluctant to leave it without some objection on that basis."

Judge Murtha looked at Tom Higgins and then back toward Melissa Ranaldo, "Well—yes, Ms. Ranaldo?"

"Your Honor," she said, "I think that we're—there is not going to be a consent order, we would like the court to rule on the merits. Obviously the court can either—if the court decides—we would like the court to do that expeditiously. The court can decide either to allow the USDA to go forward now and take the action it needs to take or it can hold this matter in abeyance, but we would prefer that the court reach the merits of that issue at this point and issue an order."

"Well, again, though, it is my understanding the government is representing that it is not going to carry out its orders until some action is taken by me."

Ms. Ranaldo agreed. "That is correct, Your Honor. And in the event that the order requested by the government is granted there would— any further relief by the plaintiffs would be moot if the court orders compliance with the order to seize and destroy the sheep."

"All right," Judge Murtha said, looking back at Tom Higgins. "I think that answers your question, or it should, that there is going to be no action by the government. And certainly if they make a representation, as they have here in open court, I suspect they will carry it out. So your fear that the government will do something else I think is probably unfounded."

"Very well, Your Honor," Tom replied.

"And certainly if they did take any action, after making that statement, there would be some consequences," Judge Murtha hastened to add.

I felt my shoulders relax. We accomplished our first goal of stopping the USDA from seizing the sheep before we had a chance to be heard in court. Larry glanced over at me, smiled, and squeezed my hand.

As he shuffled papers in front of him, Judge Murtha said, "All right. So there are a number of issues that the government has raised in response to your filings. Bearing in mind that you have not had a full opportunity to review them it does appear from my reading—quick reading—of them, that there are some questions about how much this court should be involved in the department's order.

"As you know—I think the government has indicated that because of the nature of the statute and the type of action that the government deems necessary to take, that I should not be involved only to the extent perhaps of reviewing the record of the Department of Agriculture leading to its decision or in the alternative, I think they argue that they should—or I should issue a warrant for the—to allow them to go on the land to take the sheep.

"And I have some—certainly have some questions about that procedure and how it would be followed and what rights would you have if such a warrant were issued, but these are issues that have just been raised, and which appear to me to be somewhat unique.

"I think, as stated someplace in the government's papers, that this type of action has not been taken in the last seventeen years. This is the first time since 1983 that they have chosen to proceed in this manner, so certainly we're dealing with very unique issues, and from quickly reviewing the law, there doesn't appear to be much law either.

"However, I would like to hear—perhaps what could happen is that the government could briefly indicate its argument as to why this court shouldn't proceed any further other than to review the record of the Department of Agriculture or in the alternative to issue a warrant based upon, I assume, probable cause."

Ms. Ranaldo stood. "Your Honor," she said, "we filed with the court a record—documents that were used by the secretary to reach the decision to issue the orders to dispose and the declarations in this case, and that was—that record of documents was prepared by Mr. Torres. The government has had to come up to speed rather quickly on this, and when we have referred in our motion to the affidavit of Dr. Linda Detwiler, her affidavit was prepared to summarize the administrative record and explain it in layman's terms. We have not made direct references in our brief to those parts of the record which support this. And that's primarily because that record was only received by us late yesterday afternoon . . . The record contains memoranda that was prepared for the Secretary of Agriculture. That memoranda was prepared from information of which Dr. Detwiler has personal knowledge."

"Great," I thought cynically. "Our entire court case is based on Detwiler's 'personal knowledge.'" Where is the line drawn between administrative, executive, and judicial branches of government? Could she really have this much influence?

Then Joseph Perella stood to address the judge. He was dressed in a dark three-piece suit and was another young, up-and-coming lawyer with the U.S. District Court. "Your Honor—I believe based on general principles of administrative law that are applicable here, one, our first argument here is that this is unreviewable, completely unreviewable, this substantive decision . . . the secretary's authority is described broadly. The secretary may take action if he or she determines an extraordinary emergency exists, the action that can be taken is such that the secretary deems appropriate.

"The case law interpreting such language has held that those phrases show that Congress meant to confer a considerable amount of discretion on the secretary, and in that broad discretion conferred it has essentially made it exceedingly difficult for a court to provide—to apply any legal standard because the discretion is so broad . . . There are emergency mechanisms for the Secretary of Agriculture to respond immediately, decisively to a threat that the Secretary of Agriculture in his or her discretion determines is a danger to the agriculture in this country."

The judge nodded his head and said, "And on that point I read someplace that the negotiations between the parties here have been going on for about two years. Now, what has made this such an emergency now?"

Mr. Perella coughed. "Your Honor, recently in the past couple of weeks there has been a test that has actually tested positive—a blot test that had tested positive for TSE, and that is one of the reasons that an emergency is now imminent and a decision should be forthwith.

"I would like to confine my—Ms. Ranaldo has a very extensive discussion of the need for expeditious action by the court and the basis for the agency's determination that—the factual basis for why this is an emergency, and if—with the court's indulgence I would ask to defer that to after my argument on the legal scope of review and standard, you know."

"All right," Judge Murtha replied. "Well, I think what I will do, though, is I will give the other side an opportunity to respond to your argument before we get into that aspect of it."

Perella noticeably stiffened. "Your Honor, even if the court rejects the notion that the substantive decision is completely unreviewable, at most an arbitrary and capricious standard of review applies. In any event, regardless of which standard applies, the court should be very cautious in accepting the Plaintiffs' suggestion that additional—it would be fair and just for additional procedures to be imposed on the Department of Agriculture whenever they make this finding.

"Basic principles, very well-established principles of administrative law make very clear that courts aren't free to impose additional procedures on administrative agencies that neither the agencies nor Congress have imposed themselves; *therefore, the plaintiffs do not have a right to be heard before the Department of Agriculture, they do not have a right to present an expert witness, they do not have a right to demand an independent test*" (my emphasis).

A gasp echoed through the courtroom. If we did not have the rights to any of those things, what *did* we have the right to? Was the government free to impose its will on any citizen and the citizen unable to defend themselves? Davis jumped from his seat, "Your Honor, may I be heard briefly on this, because there are some important distinctions between the Faillaces' and the Freemans' circumstances."

"All right."

"Solely on the question of the court's authority to review the agency's decision," Davis began, "we think it well recognized that this court may review standards and procedures separate and distinct from individual determinations under an arbitrary and capricious standard. We contend here that's important.

"It is most important in our case, Your Honor, because none of the Faillaces' sheep have ever tested positive for anything. If the court looks at the government's pleadings in this case, as well as the administrative record, there is no finding that the Faillaces' sheep have suffered an outbreak of a disease, have been exposed to a disease, or have a disease. And indeed the only tests that are positive were done on sheep that are not part of the Faillaces' flock and were not on the Faillaces' premises."

The judge looked confused. "Isn't the—maybe the government can address it directly, but isn't their argument that some of the sheep that the Faillaces have came from—"

"They are."

"—the Freeman place?"

"Some of the Faillaces' sheep are offspring from the Freeman sheep. The parents of the Faillaces' sheep have not tested positive for anything, and the Faillaces' sheep have not been commingled with Freeman's sheep for a period of years. 1998 or 1997 . . ." Davis said. "It is particularly evident, the problem, when you consider that the Faillaces' sheep have never tested positive yet the government declares that there is an outbreak of the disease on the Faillace farm. And if you believe the press the government would have people believe that the Faillaces' sheep have mad cow disease, when in fact they have tested negative consistently for a period of years."

The judge pondered this information for a moment and then asked, "Anybody else want to say anything on this—on these issues—?"

Ms. Ranaldo stood. "Thank you, Your Honor. Your Honor, I would like to first address some matters related to—that have been raised related to the testing. The attorney for the Faillaces indicated that—and I believe he had some discussions with the court about the developing nature of the test, and he suggested that the government should be required to use the best testing methods available.

"Your Honor, this is cutting-edge science, and—and with all due respect this is not crude testing. This is—these are tests that have been—that are the best available.

"First of all, the Western blot procedure—which is the one that was—determined the four positives in—in early July. That is a procedure that's been validated. Meaning that it has been used repeatedly, and it is commonly accepted by the scientific community as a reliable test. That was published and has been used since 1993. That test is reliable, and that is the—why the government is taking action at this point. Even though prior to this time there had been conducted—there had been tests on these sheep which indicated that these sheep were positive for TSEs, and I will get into a discussion of that in a minute. But the test that came back in early July and upon which the secretary seeks to take action now is a reliable validated test that has been used and accepted by the scientific community for—since the early '90s, and it was in use even prior to that before it became what's considered validated. Your Honor, I would also—"

"What does the test indicate there is?" Judge Murtha asked.

"The test indicates that there is a TSE of an unknown origin. Which is a—a family of diseases which causes brain damage, causes holes in the brain so that the brain looks like a sponge. So that it— the test results came back showing a TSE of unknown origin, meaning it is not something that's been in this country before, meaning we don't know what it is, or BSE. Now BSE is the disease that affects cows. This is the disease that is—was the cause of mad cow disease in England in the 19—around 1996."

"But it doesn't affect sheep?"

"The—well, we—we don't know for sure, Your Honor," Ms. Ranaldo stammered. "There is developing science on that, so that's— that's correct. There is no definitive answer with respect to whether or not sheep can get BSE; however, there—we do know that BSE has crossed a species barrier once. In 1996 BSE crossed the species barrier to cause a variant of a disease that's found in humans. It is not the normal disease that's found in humans, it was a different form of it. And it was determined, based on a paper that was prepared in 1996 by the United Kingdom spongi—the ES—the TSE advisory committee in the United Kingdom. It announced new food supply safeguards because of this variant of Creutzfeldt-Jakob Disease. . . .

"So there is—although we don't know there is no adequate science I don't think right now which would say that sheep can get BSE. We do know that BSE has crossed the species at one—at least once before. Your Honor, maybe what I can do is I would like to explain a little bit about TSEs, the transmissible spongiform encephalopathies. They are a family of diseases. They are found in animals, and they are found in humans. They affect the brains of humans and animals. They cause brain damage, and they always result in death. They are marked in the scientific world by the—a—an abnormal form of the prion protein.

"These diseases have a number of characteristics in common. First there is a prolonged incubation period. And that incubation period can last a number of years; it can last from two to five years typically.

The diseases cause a progressive debilitating neurological illness which is always fatal, there is no known vaccine or treatment for the disease, and the—the agent for the disease is transmissible. There is also no validated test for a live host, meaning what—the test has to be conducted on brain tissue from the dead animals."

"Did the people who died from mad cow disease—which as I understand it is a form of TSE?" Judge Murtha asked.

"Yes, Your Honor. It is the form that occurs typically in cows."

"Right. But as far as the humans, what did they die from?"

"They died from a variant of a similar—it is a—they also died from a neurological disease."

"Was it a variant of TSE or something else?" Murtha asked, now confused.

Ms. Ranaldo again stammered, "It was—no, Your Honor, it—they died of a TSE. They died—the human form of TSE—one of the human forms of TSE is called Creutzfeldt-Jakob Disease. Now, the brains of people who have died from this disease have been analyzed in the past. Brains of people who were suspected to have died from these since 1996, those brains—brain tissue was examined and it did not look the same as people who had previously died from this. So the scientific community has determined that it is a variant of the Creutzfeldt-Jakob Disease that killed these people, and the United Kingdom has come out with a further statement saying that this variant is believed to have resulted from the consumption of meat that was contaminated with BSE or from cows who had BSE."

"Is the Department of Agriculture worried about the potential spread of BSE or—it isn't BSE. Of spread of a variant of TSE to other animals or to humans or to both?"

Ms. Ranaldo straightened, "Well, Your Honor, the Department of Agriculture, strictly speaking, regulates livestock; however, the purpose of regulating livestock is to make sure that the food that we eat which eventually does get to humans is—"

"Well, in this case as I understand it the sheep are used—their milk is used to make cheese."

"Correct, Your Honor. And we're—we're not saying anything about the—whether or not the milk would be fit for human consumption. The problem, Your Honor, is that we have a TSE that is of an unknown foreign origin. It is a TSE, which means that it will cause this neurological brain damage which will result in death, and it is a type of—and either that or it is the BSE. Which is what cows have when it is mad cow disease.

"So the problem is that there is not a test right now specific enough to tell us that—whether or not it is BSE, and that test is two or three years away from being validated. And even—or not validated, I'm—being perfected. And even once we—if we wait two or three years until that test can determine between these two there is a—even then it may not provide conclusive results; but, Your Honor, the issue here I think is significant because it doesn't matter whether we have a TSE of unknown origin or a BSE. Either way both of these are a problem for the—the livestock in the country.

"We don't know what this TSE of unknown origin will do. We don't know how it will react, we don't know how it is transmitted, we don't know how long the incubation period is. It has only now started to show up in these cows, and these cows are four and five years old."

"You mean the sheep?"

Ms. Ranaldo's face turned bright red. "I'm sorry, the sheep. I'm sorry, Your Honor, sheep. We're only talking about sheep today and not cows.

"The attorney for the Faillaces makes the point that they—their sheep have not tested positive for this TSE of unknown origin or BSE, and that fact can be explained first of all because there is an extremely long incubation period.

"They purchased the—the progeny, the baby lambs, from the Freeman farm in 1997 and 1998. Given that it is four or five years before these sheep are even showing up positive on the Freeman farm—given that long of an incubation period you might not expect to find any positive test results on the Faillaces' farm.

"Also, Your Honor, it is significant that the—the disease—the sheep don't only have to be infected with it. All that needs—the secretary needs to do is find that they have been exposed to it . . .

"Your Honor, I know this is a complicated matter because it involves scientific facts, medical facts, tests that are in developing stages on the cutting edge of science; but what we have here is the possibility that if the government does not take action to do something given the information it has now, the positive test results with a validated test, there could be dire consequences. The—"

"What exactly is the emergency then?" Judge Murtha asked. "This has been going on for a couple of years."

"Your Honor—yes, Your Honor." Ms. Ranaldo searched for an answer. "The emergency arises now because there has been documentation and there—of—and the existence of a TSE of unknown origin. Which is different from anything else that's ever been in this country which could be a BSE.

"Now what makes this emergency is that it is extremely dangerous. We don't know all the routes by which a TSE which we know nothing about could be transmitted. We do know, however, that TSEs generally get worse over time. We know that TSEs build up in the environment making it more difficult to—to stop the spread and making it transmitted more easily, and we know that it is not easily removed from the environment.

"With a positive result, Your Honor, there is confirmation that a TSE exists, and these sheep can no longer be disposed of in ordinary means. They need to go to a special laboratory that is—it is called a bio safety level three. It has to have at least that level. There would need to be—if the USDA has these samples in its possession it would—there would need to be a special permit issued in order for these—the transportation of any of these tissues . . ."

"I assume there is nothing good about TSE?" Judge Murtha asked rhetorically.

"No, Your Honor. TSE—any TSE, any—any TSE—"

"So—"

"—causes brain damage—"

"All right. So when you say this—"

"—and results in death."

"—this is a variant of the TSE," Judge Murtha continued. "The fact that it is a TSE at all, is that significant?"

"Yes, Your Honor, It is a T—being a TSE is significant because TSEs are not—they are very dangerous, they always result in death. And in fact there is a form of TSE that has been in the United States, and that's called scrapie. Scrapie is a TSE that frequently occurs in sheep. Now, it doesn't mean that it is not as dangerous, but it—there is a certain known agent, a certain known—it is a known TSE. We can identify it, and we know what it is. It is no less dangerous because it occurs in sheep, we just can't get rid of it. That's the problem is that it exists in the United States because we can't get rid of it because it is so widespread.

"Other countries have successfully been able to get rid of it. For example, New Zealand and Australia. They—when this first became a problem they eradicated—they were successful in eradicating scrapie from their sheep, and it took drastic measures. I understand they had to kill all the sheep. But other countries, when TSEs are found, take action immediately to destroy the sheep so that there is no spread. Canada had—is also a country that has a similar approach to TSEs. Other countries are also very aggressive in dealing with them.

"Now, this is because it is—it is a serious problem regardless of what it is, so it is not—but this is even more of a problem in this country because even though we have scrapie here, this could be—this is not like typical scrapie."

"Scrapie is not TSE, then?" the judge asked.

"Scrapie is a TSE, Your Honor. I'm sorry, scrapie is—TSE is the umbrella—"

"Right."

"—and there are different kinds of TSEs that occur in different kinds of animals."

"And scrapie is one of them?"

"Scrapie is one of them."

"And that occurs—"

"Scrapie occurs in sheep; BSE occurs in cows."

"Right."

"And—"

"But there are sheep who have scrapie—" The judge was now very confused.

"There are."

"—now in the United States?"

"That's right. That's right," Ms. Ranaldo said, trying to maintain control of the conversation.

"So why aren't they trying to eradicate them?"

"Depending on the circumstances, Your Honor, it is my—it is my understanding that they are. But the problem is this—that disease is so widespread now in this country that in order to do so you have to kill all the sheep in the country virtually. In—because it spreads through the lambing and the lambing process and so it is spread from sheep to sheep. Now—"

Whispers filled the room. "Did she really just say that?" "That would mean all the sheep in the United States would have to be killed in order to eradicate scrapie!" "They are nuts!" The USDA had a program to eliminate scrapie, and it did not require killing massive numbers of sheep, let alone all of them.

"So this is worse?" the judge asked.

Ms. Ranaldo knew she had his attention now. No judge would want to be responsible for allowing the introduction of a disease with such dire consequences, an unknown disease. "This is worse, Your Honor, yes, because this is a variant. This is not like any scrapie that we have in the United States right now. It is—even if it were scrapie, it is not like it. It is a TSE of unknown origin. It might be scrapie that we have never seen before, it might be BSE, or it might be a completely different TSE that we don't even know about.

"And the problem—the problem is even with this variant of—even if it is a variant of scrapie we can't assume that it is going to have the same characteristics of the disease of scrapie that we know of in this country because we have never seen this disease in this country before. So—"

"So it is the fear of the unknown—" Judge Murtha said, mulling over what he had just heard.

"No. Well, Your Honor, it is—I wouldn't characterize it as the fear of the unknown. It is I think much more serious than that because it—we do know. It is the fear of the known. And the known is that this is a very serious disease in sheep; it causes this serious brain damage. We don't know whether or not it can jump species, we don't know whether or not or how long it may take to incubate, we don't know—but what we—what we do know is that the longer we wait the more entrenched it gets. We don't know how virulent this is; we don't know how infectious it may be. Not only to sheep but perhaps other—other species.

"We know enough about this so that not only the USDA and its experts, Dr. Detwiler, who is internationally renowned for her work in this area, but other experts in this area—not only experts, authorities, internationally agree that the USDA is taking the right steps at this point in time. Dr. Stanley Pruisner, a 1997 Nobel Laureate for his work on prion diseases, has recommended that the USDA eliminate any possibility of risk from these sheep by purchasing them and incinerating them. Other TSE researchers—"

"Are there—"

"—have also recommended this."

"Are there any that don't?"

"I'm sorry?" Ms. Ranaldo said. This was one question she did not want to answer.

"Are there any that don't, any that disagree? That you are aware of anyway?"

"I'm sorry?" she said again. She needed time, time to formulate an answer. The stakes in this case were too high. The USDA had an enviable track record, and she could not allow defeat in any form.

"Is that a unanimous opinion of the people who are involved in this field, or are there people who disagree?"

"To my knowledge, Your Honor, I am not aware of any reputable, respectable scientists who are heavily involved in the research and on the cutting edge of this area that have—that would have a different opinion. If I may have just a moment, Your Honor." Ranaldo needed some backup on this. She rushed over to the table where Detwiler and the other lawyers were seated and listened intently as Detwiler whispered to the group.

Ms. Ranaldo approached the judge, "Your Honor, I think my colleagues have pointed out to me a difficulty in presenting science in the courtroom in this matter, and I am told that sheep have been known to contract BSE, and so I stand corrected on that. The BSE has been—BSE infected products have been fed to sheep under laboratory conditions, and they have developed a TSE."

Mr. Perella corrected her, "BSE."

Ms. Ranaldo's face grew brighter as a few snickers were heard in the courtroom. "They have developed a BSE. Your Honor, I guess this is the difficulty with presenting science in the courtroom . . . Your Honor, I think what I would like to do is get back to another concern of yours, and that is with respect to why this is an emergency. Not only are the findings serious, not only does it indicate an extremely serious condition that needs to be addressed; but, Your Honor, the plaintiffs' actions in dealing with this matter have left the government with a great deal of concern.

"At this point in time I have become aware of—that there is—or I should say there was a statement in the paper that one of the Faillaces' sons admitted to chasing some rams up into the woods to try and hide them from the USDA officials. The longer these sheep are out there, the greater the risk or potential may be of perhaps one of them escaping. And if one of these sheep did escape that would have international consequences. It would impact on the pharmaceutical trade because we use animal products in developing pharmaceuticals; it would have an effect on the trade of cows, sheep, and—goats; and it could also create a potential public health scare . . ."

Now Ms. Ranaldo was going too far. We were always careful to keep the rams and ewes fenced far away from each other so there would be no accidental breeding. We also used some pasture in the old apple orchard in the summer to provide shade for the animals. Francis did not "chase them into the woods." All the sheep were still fenced, still on the property, kept according to the quarantine.

Davis stood to address the judge, "Your Honor, may I briefly? The Faillaces have brought several experts to meet with representatives of the USDA who have opposing views on the belief of the science. That should be stated for the record."

"You mean in the past two years?"

"Yeah, within the last two years. More recently than that. The Faillaces have asked that other tests be done under the supervision of experts they have consulted with, and Craig Reed of the USDA has declined those requests. With respect to the Western blot test being a validated, well-recognized, accepted test, perhaps so, but the method used can vary. The antigen that one uses to make the emulsion and to do the test. It is almost like a pap smear. You get an image on a slide, and you read the image, and you draw conclusions about the appearance of the image on the slide.

"The doctor they [USDA] have used, Dr. Rubenstein, who advocates innovative approaches, has actually published literature on the different types of results that you can get by manipulating the antigens that you use. And Dr. Detwiler has characterized the tests used in this particular case as, quote, 'new and improved.' So we don't know what the test methodology was. The test itself—"

The judge cut Davis off, "Well, I would only get into that if I were to accept the argument of a de novo review, right?"

Davis nodded. "We appreciate that, Your Honor. I am trying to set the record straight in some respects. The test does not detect disease. The fact that the tests identify the marker protein does not mean disease. The Faillaces have had tissue samples examined under more exacting scientific standards than the Western blot test, and they are disease free.

"The marker protein that this Western blot analysis identifies is the same in scrapie or BSE. You can't distinguish between mad cow disease and scrapie on the basis of this marker protein, and that's the fallacy of the government's position in this case.

"Finally, Dr. Detwiler, who is on the European commission, as set forth in the affidavit submitted by the government, has signed off on a report here that says very clearly as a general conclusion, quote, 'no single test is sufficient on its own to confirm a diagnosis of scrapie/TSE in sheep.' So for purposes of public consumption the government's scientist says one thing and for the purposes of this case they take a somewhat different position. Your Honor, the procedures here are suspect, and that is subject to judicial review."

"I have a question for you," Judge Murtha said, sitting back in his chair.

"Certainly, Your Honor."

"Since the government is willing to pay fair market value for these sheep, what's really your complaint? I mean, you are going to be made whole for the sheep."

"Well, they are irreplaceable, Judge. The Faillaces' business depends on the use of the East Friesian milking sheep because of their high milk-producing capacity; their breeding business depends on the use of the Beltex and the Charollais sheep because of what they give the breed stock. There are no comparable animals that can be obtained, period.

"For example the government argues that semen might be available from European stock that is of comparable quality. To breed the sheep from that semen in order to produce comparable stock would take many, many years. During that period of years where the sheep are bred and selected and developed into the kind of flock they need to sustain their business enterprise, of course that business enterprise would not succeed. It is an irreplaceable commodity; it is not a question of fair market value, Judge."

The judge was not convinced. "But couldn't they be replaced with sheep who maybe don't produce as much milk but there are more sheep?"

Davis shook his head. "The cheesemaking and the economies of scale for their business dictate that the number of sheep that they have be sufficient to produce the quantity of milk they need for the cheese. In other words, to support the same cheese production they would need so many other kinds of sheep that it would just not be feasible. It really is a rare commodity."

"Does that require some sort of analysis by me of the productivity of the business? Or the profitability, I should say?"

"No. I think, Your Honor, for purposes of our motion for injunctive relief the irreparable harm we contend is demonstrated simply by the fact that these animals are not replaceable, period. And I don't mean to say that this particular animal with its individual characteristics is irreplaceable, I mean the genetic stock. The contribution these particular animals can make to the business, their inherent features as this particular flock, cannot be restored. Unless, of course, we get semen and spend fifteen years breeding sheep over here.

"Finally, something came to mind about the scrapie issue. The Faillaces actually started the Vermont scrapie-monitoring program. These folks have been very involved. And for the government to contend that their sheep are somehow infected despite negative test results is really—it is difficult to accept. We do have experts, Your Honor."

The judge shifted forward in his seat, ready to end the hearing. "Well, does anybody else wish to say something?"

Mr. Perella stood. "Your Honor, we maintain our position that the burden on the Plaintiff to show arbitrary and capricious is more than just some showing there is an alternative test. They must show that the test the government utilized is arbitrary and capricious. And in order to do that they must show that the Department of Agriculture ignored some scientific information available at the time that conclusively established that this test was ineffective."

Judge Murtha called for a twenty-minute recess. When court reconvened, Judge Murtha announced that he would pursue the case further but under certain restrictions. The burden of proof would lie on our shoulders to prove the government's actions were "arbitrary and capricious," and the only matter we could argue on was Rubenstein's test and the "manner and methods he used."

Each side was allowed two witnesses, and Judge Murtha agreed to hear testimony the following week. He reiterated that he did not feel that the USDA had done anything wrong but that "it seems appropriate to hear evidence at least on the limited issue."

Discussion then took place regarding the test results, and our lawyers requested that the USDA provide better copies of the blots than the faxed copies we had in our possession. Ms. Ranaldo told Judge Murtha that the USDA would overnight photographic quality copies.

Mr. Perella stood to give Judge Murtha one last warning. "Your Honor, if it comes down to battling scientists here, so long as the government scientists have a reasonable basis for their conclusion, the court must defer to them."

17

WESTERN BLOTS AND HARD HATS

Mommy?" I heard Jackie's whispered voice. In my dream the sheep were loose. The USDA had stormed the farm with guns and was shooting them on sight. I shifted in my sleep, turning to get closer to Larry.

Strangers were in the barn trying to take the sheep, and Jackie was clinging to Mrs. Friendly. I was afraid they would take her, too.

"Mommy?" Jackie called out for me.

My heart was pounding. I had to get to her.

"Mommy, wake up!"

My hold tightened on Larry. Someone was in the room.

"Are you awake, Mommy?" Through the moonlight I saw Jackie's silhouette standing by the edge of our bed.

"Honey, what are you doing up?" I asked with a sigh of relief. It was just a dream. Jackie was here, the sheep were still here.

"I moved the rams to a new paddock and was wondering if you guys were getting up yet. Do you want a cup of tea?"

"Jackie! What time is it?" I asked. Moonlight streamed through the windows.

"About 4:00," she said quietly. "So, do you want a cup of tea? I could put the kettle on."

Larry was now awake and moved to sit up in bed. "Did I hear you say you moved the rams?" he asked.

"Yeah, it was too nice a night to sleep, and I knew they had to be moved. Isn't it pretty out?"

I laughed. It *was* a beautiful moonlit night, but the rams were at the far end of the farm, about a quarter mile away from the house. Leave it to Jackie to happily work alone in the fields at 3:00 A.M.

I loved the fact that our children had no fear of the dark. When I was growing up my sisters and I zealously attempted to scare each other. Years later we still jumped at the slightest noise, ever vigilant for the hiding boogeyman. But this wasn't the case for our children—they were fearless. They were never nervous to go to the barn alone in the dark and check on the sheep or, in this case, do chores by themselves in the middle of the night, or to be home alone.

"Why don't you let us get another hour or so of sleep, and then we'll have tea with you," I told Jackie.

She was quiet for a moment and then said, "That's okay. I'll go ahead and put the kettle on for myself. You sure you don't want a cup?"

Life returned to a peculiar sense of normalcy. I worked the store with Moe not far from my side, Francis managed the pastures, Heather milked the sheep, and, despite Jan Carney's warnings, Larry and Jackie made cheese. Sometimes hours would go by and I would forget we were under quarantine and that the government was intent on destroying our sheep and our livelihood. Those were blissful times when we were a family working together on our farm: stacking hay in preparation for the long Vermont winter, weeding the gardens, trimming sheep's hooves, and figuring out new products to add to the store.

When reality struck it struck hard. "I think we're being watched," Francis said one day.

I reassured him that as much as we were under siege, this was not a James Bond movie, and men in dark coats and glasses were not lurking around every corner.

"But I keep seeing the same cars go by when I'm working in the fields," he insisted. "They slow down and watch me. And they're cars I don't recognize."

"This is the summer, Francis," I said, "and lots of tourists are in town."

"Yeah, but these cars go by all the time. I see them almost every day. The cars have Vermont plates and are really clean. That's not a Vermonter."

I had to agree with him there. Vermonters are not known for their clean cars, because it is nearly impossible to avoid traveling on dirt roads. I promised Francis that Larry and I would keep an eye out for these "spies," but it quickly slipped my mind as we were busy getting ready for court.

We contacted Pierluigi Gambetti at Case Western Reserve University and asked if he would be our expert witness in court. He apologized and said his schedule would not allow it but recommended a coworker and professor, Dr. Shu Chen. Dr. Chen was the head of the Unit of National Prion Disease Pathology Surveillance Center, which was funded by the CDC and sponsored by the American Association of Pathologists.

Mr. Freeman hired Dr. Glenn Telling from the University of Kentucky as their witness. Dr. Telling was the principle investigator of the laboratory involved in prion research and had also worked with Dr. Stanley Pruisner for seven years. We had two of the best experts in the field.

Our next job was to learn as much as possible about Western blots. A Western blot is a laboratory procedure that separates proteins with an electrical charge according to their molecular size along a gel and allows scientists to detect and measure a single protein among other proteins. Once the charge runs through the gel, bands of various molecular weights appear along the gel. Each band represents an individual protein and is identified according to its molecular weight.

When using the Western blot to test for a TSE, an enzyme (for example, proteinase-K) is added to the mixture, which digests particular proteins in a sample and results in no bands appearing on the gel except for those resistant to enzymatic degradation. The abnormal form of the prion protein (PrPres, considered a marker for TSE) is resistant to enzymatic degradation and will appear on the band at a particular molecular weight.

The night before the court hearing Larry and I met with the three lawyers, Dr. Chen, and Dr. Telling at a small inn in New Hampshire to review Rubenstein's Western blot test results. We found mistake after mistake. The photographs of the blots—which the USDA did not give us until 6:30 that evening, despite telling Judge Murtha a week earlier they would be overnighted—clearly revealed improper sample preparation (that is, incomplete proteinase-K digestion), and the absence of negative controls and molecular weight markers made it evident immediately that Rubenstein's tests could not be relied on.

When the proteinase-K digestion step is incomplete, it results in the appearance of bands on the blot where there should be none. And the negative control—a sample from the brain of a sheep that is known to be TSE-free and which is processed and analyzed in an identical manner to all the other samples—is a protection against laboratory error, because when a blot with undigested proteinase-K was produced, the negative control and the unknown sample would both produce a false positive result.

Rubenstein ran two Western blots for testing Mr. Freeman's sheep. Each blot had ten lanes, and the first blot contained seven samples from Mr. Freeman's sheep and three positive control samples (two of which were so heavily loaded with protein that it was impossible to distinguish any banding, only a big smudge). Both blots had low molecular weight bands in every lane. On the second blot, Rubenstein re-ran samples that tested negative on the first blot and did not re-run the samples that tested positive.

The other built-in cross-references for producing Western blots are molecular weight markers. Proteins have differing weights and are identified on the blot by their molecular weight. Molecular weight markers, when run with a negative control, ensure that the normal prion protein (PrPsen) is not mistaken for the abnormal prion protein (PrPres).

Rubenstein claimed the bands on the blot demonstrated positive results for the abnormal prion protein, but without molecular weights it could not be determined exactly which proteins remained on the blot.

As the night wore on, we wondered if Rubenstein's tests were merely examples of sloppy laboratory practices, or if they represented the intentional misuse and exploitation of a complex science to mislead the public and garner support for an otherwise unjustified action.

July 27, 2000, found us in Brattleboro, Vermont, back in court for the second time in less than a week. Judge Murtha again presided, but this time the USDA did not allow Melissa Ranaldo to stutter her entire way through the hearing. This time the USDA brought out their big gun—Charles Tetzlaff.

The courtroom was almost full when we entered, but I immediately felt Tetzlaff's presence. He was a tall, distinguished-looking, gray-haired man wearing a perfectly tailored suit and a large Stetson hat. His attitude filled the room. United States Attorney for the District of Vermont, Tetzlaff would be selected for positions of authority by two U.S. presidents. His confidence unnerved me. We knew the science, but he knew the legal system. He knew how to win cases.

Tetzlaff made sure he was the first person to address the judge, and he urged Judge Murtha to limit the hearing, "and the evidence will be limited to the manner and method of testing in this case and not get into issues such as interpretation."

The USDA's opinions were spelled out in their written submissions to the court, and the degree of power to which they felt entitled was frightening: "First and foremost, there is no procedural requirement under the Act for the plaintiffs to be heard and present evidence before the order is issued. There is no requirement under that statute that the order itself provide any discussion of the basis for the findings. The fact that the Secretary has made the required findings is sufficient under the statute to support an order to seize and dispose of the sheep. Nor is there any right conferred on the livestock owner to conduct an independent test for the communicable disease . . . the inherent nature and structure of the Act of July 2, 1962, indicates that Congress left no room for judicial review."

The burden of proof was on us, not the government. We had to prove the government's actions were "arbitrary and capricious," and Judge Murtha would only look at Rubenstein's test and his methodology.

In the courtroom, Rubenstein described in great detail how all Western blots are different and that the banding pattern differs for sheep with various strains of scrapie. He used sheep and hamster samples for his positive controls: "These are processed samples that we've run previously. These are samples from known cases of scrapie sheep . . . So I was absolutely certain that those would band in the proper region because we've run those before . . ."

Yet when asked which positive control Rubenstein referred to when assessing the results, Rubenstein admitted he used the positive hamster control, not the positive sheep control, even though the hamster lane had an excess of protein that rendered it unreadable.

Rubenstein also claimed that the samples in the first two lanes on the first blot were negative, and therefore they were the "defacto negative controls." But these very same samples were listed as "±" (meaning he was uncertain of his results) on his report to the USDA. One of the reasons may have been the low molecular weight bands that appeared on the first two blots and not on the third. When questioned, Rubenstein agreed that the bands running across the bottom of all the lanes in the first two blots were some sort of cross-contamination or undigested protein. This was Rubenstein's own admission that the proteinase-K had not digested all the protein, and without molecular weight markers there was no way of determining which protein remained on the blot.

Included in the copies of the blots we were given the night before was a third blot that Rubenstein claimed was his negative control. But as Davis pointed out, there was no mention of this third blot or of using any negative controls in the lab notebooks. More important, negative controls are only valid when appearing on the same gel. According to Dr. Glenn Telling, "Each gel, each blot, is a separate experiment. There are variables between the processes of gels and the processes of blots. So you can't do an experiment on day one with controls and apply it to an experiment on day two. Nor can you do an experiment on day one and a separate experiment on the same day. It's inconsistent."

Rubenstein's third blot had four lanes (one was CWD positive, two were scrapie positive, and one was the negative control lane). Rubenstein claimed this was the negative control for Mr. Freeman's sheep and that he was testing elk and sheep on the same day and at the same time as Mr. Freeman's sheep. "The blot was run at the same time for two reasons. One was because we had these preparations we made some time ago, we wanted to check them to make sure they were going to be useful for our studies; and also I needed a lane for a negative control for this blot as well as the blot having Vermont sheep on them."

Davis asked Rubenstein if it would be sound scientific practice to have a random elk sample among sheep samples on a blot, and Rubenstein admitted it was not.

In addition, Dr. Rubenstein's procedures were invalid due to confirmatory bias.

Ms. Ranaldo asked Rubenstein in court, "So before the past week or so you didn't have any knowledge that these were suspected TSE flocks?"

"That is correct," Rubenstein replied. But Rubenstein began testing Mr. Freeman's sheep in 1998, and he was a close friend and colleague of Detwiler's—they had even published scientific papers together.

Rubenstein claimed the samples were coded and he did not know where they came from, yet every single page of the laboratory notebook had "Vermont Sheep" written on the top, and when Rubenstein reported his test results to Detwiler he wrote, "As per your request, we processed these Vermont sheep brain samples for Western blot analysis."

Another fact was Rubenstein's test was *not* validated despite the USDA's claims. The USDA loved to throw around scientific terminology, such as the word *validated*, without regard to its accuracy. Test validation is a technically challenging and scientifically rigorous set of procedures that are detailed by international standards that prove if a test used for diagnostic purposes gives true results. In order to validate a procedure, other laboratories will run the same procedure and attempt to replicate the results. Successful replication and subsequent submission and approval of the results for a scientific publication result in a test being considered "validated." Rubenstein's Western blot did not meet any of these criteria, yet the USDA repeatedly claimed in court, in their press releases, and in the documents they submitted to Glickman, which led to the Declaration of Extraordinary Emergency, that Rubenstein's test was "validated." It was another example of the USDA's manipulation and abuse of science.

When further questioned by the judge, the USDA also claimed Rubenstein's test was the "gold standard" test for TSEs. If this was the case, why did the USDA not use this wonderful test for their TSE testing program?

And how could the USDA claim what Rubenstein had found was an "atypical TSE of foreign origin"? Even Rubenstein refused to agree with the USDA's claim. And then there was the fact that Rubenstein used a polyclonal antibody, one that could detect many different proteins, versus a monoclonal, which is specific to an individual protein. "When you're doing Western blots, any lab that uses antibodies won't switch from one antibody to another antibody for every blot they do." This statement would come back to haunt him.

By the end of the afternoon it was apparent to me that Judge Murtha had his doubts about Rubenstein's test, but when Judge Murtha inquired about having the test repeated, Rubenstein claimed he had used every part of the brain that the USDA sent him and that any other section of the brain the USDA still had might not have enough PrPres accumulated in it to test positive.

"I inadvertently disposed of any remaining tissue sample," Rubenstein told Judge Murtha.

Dr. Chen was sitting beside me in the courtroom and quickly did some calculations. He quietly handed me a slip of paper. With the amount of brain tissue Rubenstein had received, he should have been able to run an additional five thousand tests. Plus the USDA had only sent him a third of the brain tissue, and two-thirds should have still been in their possession.

In his closing argument Charles Tetzlaff said that the USDA was given broad powers by Congress and that their responsibility was to protect animal and human health in the United States, and they were justified in claiming an "extraordinary emergency" even if the animals only had scrapie because their ancestors originated in Europe. As explained in their documents for the court: "Significantly, four of the sheep that tested positive on the blood test likewise tested positive on the Western blot test, the validated and accepted test, conducted by Dr. Rubenstein. Thus, two tests have confirmed the presence of an atypical form of scrapie of an unknown origin or BSE in these sheep even though they were certified to have originated from scrapie-free flocks."

Both Drs. Chen and Telling questioned the results and said, as their expert opinion, that the tests should be re-run. But neither scientist would slam Rubenstein and his test in court, because the world of science is much like the world of politics and you have to be careful not to burn bridges. The TSE universe is a tiny blip in the cosmos of science, and all the scientists know each other and often collaborate with one another. Insulting or criticizing the performance of another scientist is not acceptable and could cost you your career. In fact, unbeknownst to us at the time, the day prior to our court hearing Dr. Rubenstein and Dr. Gambetti (Dr. Chen's boss) met to discuss future collaborations.

When the hearing finally ended, Judge Murtha said he would issue a written decision as soon as possible.

I'm not sure why, but it always seemed that whenever life was busiest, we moved. Our move to Roxbury Mountain Road in the midst of the battle was a blur and was only accomplished with the incredible assistance of many friends who came from as far away as Burlington to help. Then Pélé came home. Members of the media had worried about his safety and snuck him off the farm the morning of our press conference. For the next few weeks Pélé had an underground railroad of homes in the valley. Francis was overjoyed.

Life was just starting to settle when I received a disturbing phone call from a Michigan shepherd we had met at the Wisconsin sheep conferences.

"Linda, you need to file a FOIA for test results," he told me.

"A what?" I asked.

"A FOIA—Freedom of Information Act—it allows you to request data about the activities of federal agencies, like the USDA."

"And why would I want to file a FOIA?"

"Because there are more tests that you don't know about. The USDA has kept some hidden from you."

"No," I argued. "We have all the test results from the USDA."

"No, you don't," he countered.

Again, I disagreed with him and explained that we had copies of all the results from our animals that were tested. I appreciated him taking the time to contact us. Phone calls, letters, and e-mails were coming in every day, and everyone had an idea that would save our sheep: give a single sheep to a hundred different homes in the valley, making it impossible for the USDA to find them all. Sounded great, but taking even one sheep off the property would break the quarantine and allow the USDA to immediately seize the animals and not be obligated to compensate us.

"Get PETA at your farm. They'll fight off the bastards if they try to take your sheep," a fervent sheep lover from Massachusetts told me—repeatedly.

Others suggested moving them in the middle of the night across the Canadian border, or, one of my favorites, donate them to the Belgian royal family. The strangest show of support came one night around 10:00. There was a knock on the door, and in walked a middle-aged man wearing a bright orange construction hat. He warmly shook our hands and said he had been following our story closely and wanted to help.

"Do you think you're being monitored?" he asked. We told him about the strange noises on the phone and that Francis was suspicious of some of the people driving by. The man asked for more details about the phone and after listening intently to our answers shook his head sadly and said, "Yep. Yep. You're being tapped."

It sent chills up my spine when he said it. I tried to shake it off. Why would anyone be interested in surveilling us?

"I want to help you, to save your family," he said. "First you need to unplug the phone from the wall."

I looked at him quizzically.

"You didn't know they can listen to conversations you are having, here in your home, even if the phone is just on the cradle?"

I shook my head no and unplugged the phone while he told us about his feud with the government. The more details he gave, the more similarities I noticed to our situation.

"They're not going to stop," he said. "Not going to stop until they put you out of business." This struck a little too close to home as I thought of what Wayne said when he gave us the quarantine papers. Instinctively I moved closer to Larry.

"And after they destroy your business, they will go after your family and try to tear that apart. You have to be prepared."

Larry squeezed my hand reassuringly. He could tell how much this conversation was bothering me.

"You need to be strong. You need to be a team. You have a wonderful family, and working together you might just save yourselves," the man warned.

We thanked him for taking his time to come to our home and talk with us. As he was leaving Larry asked him about his hard hat, which had remained on his head the entire time.

"That's to protect me from them," he said cryptically.

"Them?" Larry asked.

"The government." He explained how government officials had managed to put a probe in his brain so they could read his thoughts.

"The hard hat screws up their signals," he said.

And so it was with this in mind that I assured the fellow shepherd from Michigan that the USDA was not involved in a conspiracy and were not hiding test results like villains in an old spy movie.

"You need to trust me on this one, Linda," the shepherd pleaded. "I know there are more test results. Would you please just file the FOIA?"

I didn't know the first thing about "filing a FOIA" and had more than enough on my plate as it was. The shepherd heard the pause and interpreted it as a positive response.

"Okay. Great. I'll e-mail you the FOIA letter. Just print it out, sign it, and let me know when you get a response."

Wearily I agreed and thanked him again for his time and effort. I realized his intentions were good, as were those of the man in the hard hat.

But what I didn't expect came fourteen days later.

18

USDA RENEGES

Judge Murtha found in favor of the USDA on August 1, 2000, and denied our request for a preliminary injunction to stop them from seizing our animals. We were crushed as we read through his ruling. Judge Murtha was misled into thinking the four animals that tested positive by Mary Jo Schmerr's experimental blood test were the same four that tested positive by Rubenstein's test and therefore considered Rubenstein's test confirmatory. But the animal that tested "most positive" on Rubenstein's test was negative on Schmerr's. Judge Murtha thought the USDA had been very reasonable by offering us compensation and sourcing genetics. What he didn't know was that the USDA had still never given us a specific financial offer. They refused to say anything more than that we would receive "fair market value" for our animals and that a final number would be determined *after* we surrendered our sheep. The source of East Friesian genetics the USDA located was not even close in quality, blood lines, milk records, or production to ours. The Beltex were irreplaceable. We had the only Beltex sheep outside of Belgium, the Netherlands, and the United Kingdom, and because the doors for importation had been shut, there was absolutely no opportunity to replace them.

Murtha wrote, "If the secretary is correct that the continued presence in this country of the Belgian sheep constitutes an unacceptable risk, their destruction will greatly reduce the possibility that humans and domestic animals will be exposed to a contagion not presently endemic in the United States. Such potential harm, together with the government's obligation to compensate the plaintiffs for their loss, counsel against the granting of injunctive relief."

Judge Murtha gave us six days to appeal. We did, and on August 4th Judge Murtha ruled in *our* favor. "The plaintiffs' motion for certification to file an interlocutory appeal is granted. Under the exceptional circumstances present in these cases, the failure to permit the plaintiffs to appeal immediately will result in their inability to halt the USDA's action, and therefore to obtain the relief they seek on appeal."

We did not realize the USDA was desperate to avoid second circuit court, and so when they quickly offered an alternative—if we gave up our appeal, the USDA would allow Murtha to expand the case—we agreed.

That very same day we received the best news of all: The sheep were going home! Belgium was granted permission by the European Union to repatriate our sheep. The Belgian Minister of Agriculture, Jaak Gabriels, sent a letter to Secretary Glickman requesting the animals be returned to Belgium.

Immediately we met with Mr. Freeman to make travel arrangements for the sheep. Mr. Freeman offered to pay for flying the animals and hire a DC-10, while Larry and I would order shipping crates. It was not the way I ever imagined our sheep farming life would turn out—import sheep from Belgium and four years later export them back—but the smiles on the children's faces and in their eyes were worth it.

While we were making plans to send the sheep home, Francis became more and more suspicious of the USDA. Unable to get support for his theory we were being watched, he went to Bruce. When Bruce was working the store one summer's day, Francis called him from the pay phone on the front of the schoolhouse, thirty feet away.

"I've got an idea," Francis whispered.

"What?" Bruce asked.

"Meet me at 1:00 A.M. I know a place where we can hide the sheep."

"Great! I'll see you then. And Francis?"

"Yeah?"

"Don't forget Moe."

The vehicles driving around the farm that night in search of the fictitious sheep kidnapping were enough for Larry and me to realize that Francis was right. Someone *was* listening and watching us. The top of Roxbury Mountain Road had an expansive vista of the valley and a full view of our entire farm. Day after day, customers and friends reported that two men in dark glasses parked for hours at the lookout. Strange people were seen climbing in and out of the paddocks with the sheep. (Why anyone would enter into a pen of rams during breeding season is beyond me.) Then one morning Francis awoke to find two men parked fifteen feet away from his bedroom window, peering inside the house to see if anyone was home.

Who were these men, and what were they doing? What did they want? I struggled with this new information. Did I need to worry for the safety of my family? I didn't want to instill fear in the children, yet what were these men after? It was a fine line to walk, and I decided to treat it with humor. When Larry and I talked with the children about the surveillance I told them to think of it "like a Monty Python movie," because the agents watching us were not very good at staying hidden. We began giving suggestions when we heard the agents on our phone: "You really should try the glider rides at the Warren Airport. Some of the best gliding in the country is here in the Mad River Valley. Plus you get a great view of our farm. And then you should have dinner at American Flatbread. It takes a long time to get seated, but it really is worth the wait."

But behind my smiling mask was a growing unease. What if they did something to the sheep? The sheep were in various paddocks across the farm, and most of them could not be seen from the house. Without being asked, friends set up tents and camped in the fields in order to be near the sheep at night. A close friend and supporter, Barry Simpson, loaned us a floodlight on a twenty-foot pole that he used in the winter for illuminating the ice-skating pond. Our family gathered all the sheep and put them behind the schoolhouse so we could see them from our bedroom window. The bed in our apartment was on a wooden platform, five feet off the floor, and flush with a bank of windows. I lay awake night after night, vigilant for any unfamiliar shadows around the animals.

Larry and I told Davis we thought we were being monitored. When he mentioned this to the USDA they told him, "Your clients are paranoid."

Ken Squier, owner of the local radio station (WDEV), had an idea. "Get some license plate numbers," he instructed Larry and me.

Bill Carnright and Edie Connellee were guardian angels disguised as home inspectors. In their midfifties, they lived a full life and always followed their passions. We first met them when they inspected our house. From day one, Edie was passionate about the abuses we were suffering at the hands of the USDA. "My brother and my cousin in

Texas have been following your story and are as outraged as we are at what the USDA has done to you," she told me as she examined the kitchen, peering into cabinets, checking the plumbing under the sink. "I'd like to give the USDA a piece of my mind!" We quickly became friends, and whenever things were tough I always knew I could turn to Edie for support.

I told Edie about the sightings, and she immediately outfitted the entire family with walkie-talkies, cell phones, and cameras that were hidden in our pockets and baskets as we worked around the farm. One morning Larry and I had just stepped outside our house when Greg Hughes (Wayne Zeilenga's assistant) drove past in a white USDA vehicle. At the same time Francis and Jackie came running from opposite ends of the farm. Francis saw federal agents in a gold Mitsubishi Gallant watching him as he moved the rams. Jackie noticed two men, one of them holding a black box, climb into a lamb paddock while she and Heather milked the sheep. Larry then spotted the Mitsubishi going past the four corners, and we attempted to race after them but could not catch them.

An hour later Edie called. "They're here! They're here!" she shouted into the phone.

"Who's where?" I asked.

"The USDA. They're down in Warren Village, and now they're turning onto Brook Road, and they're haulin' ass!"

Larry and I rushed to our truck and just as we were ready to pull out of our drive, the Mitsubishi passed us. Unaware *they* were being stalked, we followed them to the top of Roxbury Mountain where they turned around and began the descent back down the mountain. I rolled my window down and flagged the two men as they approached.

Reluctantly they slowed and lowered their windows.

"Can I help you?" I asked, smiling.

They didn't smile back as they realized who we were. Finally the driver shook his head in annoyance. "No," he said, and continued back down the hill.

As quickly as possible I turned the truck around and raced after them.

"We forgot the cameras and the phones," Larry groaned. "And I can't see the license plate numbers."

I sped up to try to reach them, but the faster I went, the faster they drove. "They'll have to stop at the four corners," I said, referring to the intersection of roads near the Schoolhouse Market. "Hopefully we can catch up to them there."

When we topped the last hill before the stop sign, I saw Bill and Edie at the end of the schoolhouse parking lot. They would have their cameras. I pushed the pedal down as far as I dared and laid on the horn, pointing at the car ahead of us.

Just at that moment, Pélé noticed some people walking near the intersection by the schoolhouse and ran right in front of the car. The agents slammed on their brakes as Pélé gleefully jogged past. I came to a screeching halt, inches from their bumper. The federal agents were trapped, with the walkers and Pélé now blocking the intersection in front of them and Larry and me behind them. Bill, Edie, and the children rushed toward the car, snapping photos as the agents slunk in their seats.

The headline story on the WDEV noon news was a report about how a Mitsubishi with Vermont license plate number CLX114 and a USDA vehicle with license plate number A283946 were seen patrolling the Faillace farm. Within twenty minutes Davis received a call from the USDA.

"The gold Mitsubishi was not us."

"What do you mean it 'was not us'?"

"That was the office of the inspector general."

"What?!"

"There are three different groups monitoring the Faillaces right now, and the gold Mitsubishi was not us."

"*Who* is watching them?" Davis asked in disbelief.

"The office of the inspector general, the investigative and enforcement services, and the U.S. marshal service."

That afternoon, the Mitsubishi returned. This time Larry and Jackie followed them over Roxbury Mountain and toward Montpelier. The agents turned left onto Route 302 and then started to turn right into Montpelier but quickly changed their minds. When they neared the entrance for Interstate 89, they began to head south and then swerved at the last minute to go north. Larry veered quickly and managed to keep up with them. By the time they passed the next exit the truck speedometer was reading 85 miles per hour. Larry and Jackie decided to abandon the chase and come home.

• • •

The harassment didn't stop there. USDA lawyers gave Davis copies of the USDA appraisals along with a criminal investigation of Larry, which Detwiler had requested in order to find out if there was anything illegal with the importations. "Your husband is clean," Davis teased as he handed me the documents.

"I don't believe this," I said with shock as I flipped through the forty-two pages. "How is it possible that someone like Detwiler, with her midlevel ranking, could order a criminal investigation of a private citizen?" Davis shook his head and said he would do some of his own research.

I looked through the report, which was put together between November 1999 and February 2000 by the Investigative and Enforcement Services of the USDA, "one of the groups that is watching us," I reminded Larry sardonically. The cover page said "Violator: Lawrence S. Faillace" and gave our address, phone number, and his social security number. The next page was the following synopsis:

Violator: Lawrence Faillace is the President of Ag-Innovations, Inc., a for profit organization with expertise in animal reproductive physiology, including artificial insemination and embryo transfer. Faillace possesses a doctorate degree in Animal Science from Virginia Tech.

Business Information: Faillace operates a sheep dairy farm. His objective is to establish a dairy sheep industry in Vermont based on the production and sale of quality cheese. He also wants to establish an education and demonstration center dedicated to the opportunities of sheep dairying for Vermont farmers. Faillace sells cheese and eventually intends to sell breeding stock from his East Friesian flock.

Primary Witness: Dr. Roger Perkins, Senior Staff Veterinarian for Veterinary Services, National Center for Import/Export, corresponded with Faillace in advance of the importation of sheep, and he reviewed, assessed, and approved applications for importation.

Previous History: None in IES tracking system.

Violation Events: On July 2, 1996, and October 19, 1996, Dr. Faillace was responsible for importing two shipments of sheep from Belgium into the US. Forty-two (42) head of sheep in the first shipment and six (6) sheep in the second shipment originated in flocks in the Netherlands. At the time the Netherlands did not participate in a program determined by the APHIS administrator to be equivalent to the Voluntary Scrapie Flock Certification Program in the US. The imported sheep were accompanied by certificates issued and endorsed by Belgium official veterinarians only, and the certificates did not contain the required certification statement listed under 9 CFR92.435 (g).

"What a load of hogwash!" I said angrily. "This is what happens when you have a monstrous bureaucracy with more than eighty thousand employees. One hand doesn't know what the other is doing." The USDA gave us authorization to import the sheep from the Netherlands and quarantine them in Belgium. Every single sheep came from a flock enrolled in the scrapie surveillance program. Worse still, it was *not* a requirement of the importation for the sheep to even be in a scrapie-monitoring program. Detwiler had wasted taxpayer's money.

The list of injustices continued. Two weeks after I mailed the letter, the FOIA revealed *four hundred* negative results on sheep from our farm and from Freeman's, which we had never seen—including additional negative tests on the four sheep that Rubenstein and the USDA claimed were positive. Four hundred negative test results! I was stunned. How could the USDA hide test results from us? Detwiler assured us we were working together for public safety and "the betterment of science." What else were they hiding? We soon found out.

When we showed Davis the negative test results, he demanded the court force the USDA to surrender their entire file. "I want to know who at the USDA made the affirmative decision not to consider these results," he said. "And I want it to be shared with the court." Yet the USDA lawyers said they would only send a letter to Judge Murtha. This infuriated Davis even more. "We filed a motion demanding the government come forward with information we knew they had, information that was not provided to the secretary of agriculture and was not provided to the court," he wrote in a letter. "Rather than formally respond to the motion, the U.S. attorney wrote a letter acknowledging the existence of all these other materials and proposing they make it part of the record. That to us gives rise to the additional concern as to what else is out there."

Davis submitted the feed records to the court, and the USDA conceded that the feed records had also *not* been shared with Secretary Glickman. When questioned about the feed records, Detwiler replied, "It was all considered, trust me. I'm not sure every document the department has was sent up to the secretary. Things were summarized."

And then the Whittens finally identified the animals that the USDA alleged were positive and discovered they were thirteen-month-old ram lambs. Unbelievable! These animals were too young to test positive for a TSE—even if they were infected. (The youngest confirmed case of scrapie in a sheep was eighteen months old, and the youngest confirmed case of BSE was in a twenty-month-old bovine.)

The USDA had claimed one of the four animals that tested positive was an original import (that was their justification for attempting to seize all our sheep), and this lie was spread by press releases from the American Sheep Industry Association and the Vermont Sheep Breeders Association. The USDA lawyers implied to Judge Murtha that the animals that tested positive were four and five years old, not thirteen months.

The USDA finally realized they were on shaky ground. At a status conference with Judge Murtha in early September the USDA asked for permission to supplement the administrative record.

Judge Murtha was intrigued. The USDA had already submitted their seventeen-page administrative record, which they claimed Glickman based his decisions on.

"And exactly how large is this 'supplement'?" Judge Murtha asked.

The USDA lawyer looked at the ground and mumbled.

"Could you repeat that?" Judge Murtha asked. "I couldn't hear you."

"About one thousand pages," the lawyer said, looking very uncomfortable.

"Did you say one thousand pages?" Judge Murtha asked incredulously.

"Yes, Your Honor," was all the lawyer could say.

Davis turned and winked to Larry and me. This would be interesting.

Larry and I spent hours in Davis's conference room poring over the 1,116 pages of documents, and even more pieces of the puzzle fell into place. In June of 1999 Mr. Freeman gave sixty-five animals (including the infamous four) from his flock of more than 250 to the USDA along with a proposal letter offering all the remaining original imports. He suggested that if the animals were killed and all tested negative, the USDA would only quarantine the farm for an additional two years, at which time there would be enough evidence that all the sheep were healthy and not a risk. Even Senator Leahy sent a letter to Secretary Glickman in support. "My office has received correspondence on behalf of the farm's owner outlining a proposal which makes considerable strides toward resolving the situation. While it may not meet the Department's preferred option, it does represent serious progress. For this reason, I would hope the Department would not reject the proposal out of hand, but would consider it as a basis for further negotiation."

On November 1, 1999, Alfonso Torres sent a letter to Mr. Freeman acknowledging receipt of the sixty-five lambs and of his proposal and said thanks, but no thanks. Torres also noted, "Per your proposal, we have purchased the rams that constitute this year's progeny, though they do not enhance our ability to evaluate the risk of TSE in this flock due to their young age." Even then the USDA knew the animals were too young to test for TSE.

According to the USDA website, "USDA/APHIS has an active surveillance program in place and actively shares information and coordinates closely with other Federal agencies . . . in order to ensure that the US has a uniform approach to transmissible spongiform encephalopathies which is based on sound scientific information. APHIS' surveillance program is based on laboratories histopathologically examining brains from high risk cattle (i.e., cattle that exhibit clinical signs of a neurological disorder). If the sample is negative [by histopathology], no further testing will be conducted at APHIS' National Veterinary Services Laboratory (NVSL), the national BSE reference laboratory. An inconclusive sample would undergo confirmatory testing using immunohistochemistry (IHC) which is recognized internationally as the gold standard for BSE testing . . ."

IHC *was* the international "gold standard" for TSE testing (including scrapie, BSE, and CWD). So why did USDA lawyers swear in court that Rubenstein's Western blot was the "gold standard"?

If American cattle are negative by histopathology, nothing else is done, but if European sheep test negative, the USDA goes to extraordinary measures to keep testing, attempting to find a positive, and if all else fails, the USDA will concoct positive test results in order to justify their actions.

"Based on what we know about the transmission of BSE, APHIS would not be depopulating an entire herd," the website assured farmers. So why would the USDA need to seize all our animals?

Detwiler repeatedly bragged that a cumulative 10,499 cattle brains had been tested in the United States between 1990 and April 2000, and "no evidence of either condition has been detected by histopathology or immunohistochemistry." What she did not tell anyone was all our sheep were also negative by these same tests—every single one.

In May of 2000 Detwiler, Rubenstein, and a colleague, Elizabeth Williams, wrote a paper about TSEs for the journal *Emerging Diseases of Animals*: "The diagnosis of BSE is based on the occurrence of clinical signs of the disease and currently is confirmed by postmortem histopathological examination of brain tissue." None of Mr. Freeman's (or our) animals ever exhibited clinical symptoms. The paper continued:

> Studies conducted in the Netherlands and the United States indicate that IHC appears to be useful in detecting scrapie in preclinical sheep. This research has revealed the presence of PrPres in the tonsils of preclinical sheep (Schreuder et al., 1996) and lymphoid tissue of the third eyelid (O'Rourke et al., 1998). And, according to *Cell Biology* (September 1999), "PrPres always accumulates prior to cellular changes in the brain."

From the four hundred test results we ascertained that the USDA killed Mr. Freeman's lambs in June of 1999 and sent the brains to NVSL, where they were tested using histopathology and IHC. NVSL tested tissue from the exact same area of the brain that Rubenstein did. All animals tested negative.

The third eyelids were extracted and sent to Dr. Katharine O'Rourke, who tested all of them in August of 1999. Every single one was negative. And if the USDA was concerned that the animals had abnormal prions and were in the early stages of a disease, they would have been detected by Dr. O'Rourke's IHC, or by the IHC performed by the National Veterinary Services Laboratory in Ames, Iowa. When all this testing was completed, the remaining tissue was then thrown in a freezer where it began to autolyze (enzymes within the cells begin the process of self-digestion).

Eleven months later, blood from the same animals was tested with Dr. Schmerr's experimental test, a test she asserted was not ready and gave unreliable results that would never hold up to being challenged in court because the positive control tested negative. Yet after reviewing our legal papers, Dr. Mark Hall, a veterinary pathologist working for the USDA, told USDA officials, "Although Dr. Schmerr's test is not considered an 'official diagnostic test' at this time by the USDA, the detection of abnormal PrP by this test certainly must be considered as additional evidence."

Finally, when we refused to give in to the USDA's demands to hand over the sheep, they sent the freezer-burned samples (which were now a year old) to Rubenstein, who contrived four false positives using extremely shoddy testing and then claimed he could not run any further tests because he "inadvertently disposed of any remaining tissue samples."

Sound scientific practices? No. This reeked of incompetence, or worse, manipulation.

Our world came crashing down on August 24, 2000, when Glickman said no, the sheep could not return to Belgium. Despite the fact that the USDA's Alfonso Torres sent a letter to the Belgian Embassy on November 18, 1999, asking if they would allow our sheep to be repatriated, and the APHIS TSE Working Group recommended "re-exporting the sheep and their progeny to Belgium," Glickman refused. His reasoning? Returning the sheep to Belgium would "undermine confidence in the integrity of the [U.S.] animal health system."

We were devastated. The USDA's main concern was how the U.S. trading partners might react to the possibility of mad cow disease in the United States. When interviewed by the *Boston Globe* about how our situation impacted the cattle industry's ability to export more cattle to Europe, Detwiler admitted it would be greatly enhanced if our sheep were destroyed. "Being able to say there is no hint of mad cow disease in U.S. livestock is a big selling point. It would be in a lot of people's interest," she said.

The USDA had thought they could contrive some test data and we would roll over and surrender the animals. All farmers who had previously attempted to fight the USDA were quickly and thoroughly squashed, and with the arrogance of a bully who won all his battles, the USDA went after us. They never imagined we would end up in court, which gave us time to uncover the additional test results and correctly identify the animals. None of this was supposed to happen. The USDA was trapped in their web of deceit.

19

THE BUYOUT

N ext on *Switchboard* this evening, we will have Senator Leahy in our studio . . ." Larry leaned over to turn up the radio. "This should be interesting," he said. We were on our way to Montpelier to watch Francis in a soccer game, and Senator Leahy was on Vermont Public Radio talking about DNA testing of inmates in order to prove their innocence or guilt.

"Why don't you pull over there?" I suggested, pointing to an area beside the soccer field where we could watch the game from the vehicle and still listen to the radio. Larry parked the truck and turned up the volume.

"Our guest is Senator Patrick Leahy. Let's go to our call. We start off with John, who is calling from Warren. Hi, John, welcome to Switchboard."

"Thank you. Good evening, Senator." Larry and I smiled at each other, immediately identifying our friend John Barkhausen as the caller.

"Good evening," said Senator Leahy.

"I fully support you that our death row inmates should have the right to DNA testing. But I heard on an NPR [National Public Radio] show the other day that prosecutors were arguing that it would undermine confidence in our judicial system, and it struck me that this is the same logic—the undermining of public confidence—used by Agriculture Secretary Glickman in refusing to return the Belgian sheep to Belgium. This is even though Belgium wants them back, and we want to get rid of them, and it would provide a quick resolution."

John continued, "It seems that in both cases we are more interested in killing something or someone and making a big show than in establishing the truth. And this kind of reasoning alone undermines my confidence in the judicial and animal health systems. And I can only think that Secretary Glickman wants to kill these sheep to destroy the evidence and the USDA's shoddy, unvalidated testing procedures, just as prosecutors around the country are now destroying evidence to avoid bringing up all these old trials and having to retry them. When government officials argue that the truth will undermine public confidence in the government, I feel like there is something really wrong with the government. There can be no justice without truth, and there is no room in a democracy for this kind of totalitarian, bureaucratic thinking. I was hoping that you could bring your influence to bear on your friends and connections with USDA to change this decision on the sheep."

"Wow, that was well worded," I told Larry.

"Shh! I want to hear Leahy's response," he said.

"Well, for one thing," Senator Leahy said, "their concern I think, Secretary Glickman on his behalf, I would say this: He is concerned that the Belgians have not provided all the documentation regarding the origin of these sheep. That they have held back documentation on the potential exposure to contaminated feed of the original imports from Belgium. And that his concern is that the Belgians want the sheep back so that they can cover up things that they've held back in the first place. I have no idea whether they did or not.

"What I have suggested is that whatever is done in the further testing of these animals, that if the Belgians are concerned, that they be given full access to all the testing. That they have their experts come over and follow that. And that we make public, the U.S. makes public, each step and all the documentation—each step as they go along.

"There is this feeling of many, as I said, that the Belgians have held back information to begin with and had that information been there, some feel that the sheep never would have been brought into the United States. I know that it was never shown to me, and I was the one that wanted and hoped these sheep could come to the United States. I thought that it might be good to start a new type of agriculture, a diversity of agriculture, here in Vermont. But I am concerned when I find out that the Belgians held material back.

"I have gotten a huge number of calls and letters from people in the agricultural community who are concerned that their own work will be tainted by what's going on here. But I have a great deal of trust in Judge Murtha, who has this before his court. I think people on both sides agree that he is a man of not only great legal ability but total integrity. And I trust what he would do. I am not going to try to get into influencing either his final decisions or the appellate decisions, which, of course, I could not influence."

"John, thanks for your phone call," said the moderator.

"What the heck?!" I yelled at Larry. "How can he say these things? Belgium has provided *all* the information! This is crazy!"

"Maybe you should call in," Larry suggested.

"All right. I will. Where's a phone?" I said to Larry's surprise.

"Over there." He pointed to a pay phone about fifteen yards away. "Remember the number is 1-800-639-2211," he called after me.

I rushed over to the phone and dialed. It was busy. I tried again. Still busy. Larry questioned me with the thumbs up sign. I shook my head no. I tried again. On the fifth time I got through. I gave Larry the positive sign as my heart was pounding.

"Hello, this is *Switchboard*. What is your name, where are you calling from, and what is your question?"

"Hi, this is Linda, and I'm calling from Montpelier," I said, instinctively feeling if I said I was from Warren they would not put me on the air.

"And what is your question?" the operator asked jovially.

"I would like to speak to Senator Leahy about the sheep."

"I'm sorry. We are not talking about that topic today."

"But he just spoke about it with the last caller," I said.

"I'm sorry," the operator repeated in a friendly tone. "If you like I can give you his office number, and you can contact him directly."

"Look," I said, frustration creeping into my voice, "this is our flock of sheep, and Senator Leahy said things that were false about them."

"You will have to contact his office directly if you have a problem," the operator replied, all traces of amiability now gone.

"But what he said was wrong. People across three states and Canada are listening to this program, and he lied about Belgium and our sheep."

I crossed the line with that statement. "I'm sorry," the operator angrily replied, "but if you have an issue with the senator I suggest you contact his office directly. Would you like his number?"

After years of being pushed around by the USDA, wanting our elected officials to protect us, having our phones tapped and federal agents swarming our farm, providing the USDA with every document they requested, trying to work with governments and getting caught in sleazy politics, I had had enough. "*Switchboard* is not the public forum it pretends to be," I said quietly, "and I think we will hold a press conference tomorrow to let this be known."

There was silence on the other end. The operator was not used to being threatened. "Please hold," he said.

"Our guest this evening is Senator Patrick Leahy, and our phone number is 1-800-639-2211 if you've got a question or a comment for the senator. That's 1-800-639-2211.

"Let's talk to Linda, who is calling from Montpelier. Hi, Linda, welcome to *Switchboard*," the moderator said in a singsong voice.

I took a deep breath, "Thank you. Good evening, Senator Leahy. This is Linda Faillace, and I was very dismayed when I heard your comments about these documents from Belgium not being produced. Because I believe that you would remember on March 25, 1999, my husband and I brought over three European experts from Belgium and the Netherlands to meet, not only with Michele Barrett who is an aide of yours but also with the USDA, in the USDA's office in Washington. And the Europeans produced all the documentation and offered all of it to the USDA. And at that point the USDA said, 'There's absolutely nothing wrong with your sheep and no, we do not need the documents.'

"Later in October of 1999, Linda Detwiler claimed, in front of Jan Carney [Vermont Commissioner of Health], that these documents had not been offered, and so they requested the documents from Belgium. On November 16, 1999, all those documents were given, hand-delivered to the American Embassy, and then received on Linda Detwiler's desk in November 1999. And I have copies of all those documents, and I would be more than happy to share them with you."

Senator Leahy was silent. This was not what he had been told. Did I know something he didn't? There was dead air on the radio. Who was telling the truth? "Well, I'll raise that point with the Department of Agriculture," Senator Leahy finally said. "As you know, I have met with him [Glickman] several times on this. You also know from the time when we met and the times when I met with the Freemans, I have tried . . . I was one of the ones who supported bringing these sheep over. These documents, if they are available, should be in the court. I would hope that they would go there. But if there are documents that they have not received, I think your attorney should make sure the court is aware of that. Should make sure that Secretary Glickman should be aware of that. What I have said continuously is that whatever testing is done and all should be as transparent as possible, open and available not only to you, but to the American public, and to the Belgians."

"Linda, thanks for your phone call."

By the time we returned home from the game, I was so wound up that I gathered all the USDA files from the office and spread them across the dining room table.

"What are you looking for?" Larry asked, bemused. He enjoyed watching me fight back rather than feel overwhelmed and depressed.

"I want to find out what happened to all those documents," I said, clearing space on the counter for more files. "See if I can trace where they went."

I organized all the papers by date and spent the next three hours reading. The story began to emerge: In July of 1998, Wayne Zeilenga was concerned about granting us our certified scrapie status. Wayne sent a letter to Bill Smith asking for more information about our flocks and, despite the fact that his superiors in the USDA authorized our position in the scrapie program, Wayne asked permission to deny us certified status. It was granted.

Bill Smith then sent a letter to Larry and me and requested specific information about the flocks of origin, which we forwarded to Dr. Carton. On September 18, 1998, Dr. Carton sent the required Belgian information to Larry and me, and we forwarded it to Detwiler. Detwiler e-mailed Carton and thanked him.

At the meeting in Washington the following March, Dr. Carton offered copies of feed records from all the source flocks, but everyone refused. Yet for months following, the USDA and Detwiler denied the documents even existed. Angry and frustrated over the accusations of not providing information about the flocks, the Belgian Ministry of Agriculture hand-delivered the papers to the American Embassy on November 16, 1999. The papers were sent to Detwiler.

The USDA then requested additional information regarding the sheep and the importation, and on December 16, 1999, Relindis Joosten of Belgium sent all the requested data—again to Detwiler. The request for information kept going.

In May of 2000, Detwiler sent an extensive list of questions almost identical to those she asked of Dr. Carton to Dr. Braum Schreuder, a TSE researcher from the Netherlands. He responded within five days. Detwiler now had all the European information in her hands, yet she continued claiming that Belgium had never responded to the USDA's requests for information. Why?

Meanwhile, the mastitic ewe that Mr. Freeman gave the USDA in 1998 was tested, and tested, and tested. Detwiler dispatched tissues to everyone under the sun, and, when all was said and done, nine tests (including a Western blot by Rubenstein) were run on that particular ewe. Detwiler even wrote again to Dr. Braum Schreuder, this time to ask if there had ever been a documented case of scrapie in an East Friesian. Dr. Schreuder apologetically said the only report he ever heard was in one crossbred. He wished Detwiler luck and urged her to "keep clutching at straws. Hope you've got your own supply."

Every single test on Mr. Freeman's ewe was negative. Yet Detwiler requested and was authorized to send samples from the ewe for a mouse bioassay. The mouse bioassay is a test used to determine which *form* of TSE an animal has (for example, scrapie as opposed to BSE). The procedure takes two years, costs more than $30,000 *per* sample, and is only performed on positive samples.

Why would Detwiler and the USDA waste taxpayer's money testing an animal that was already negative nine times? Because the USDA needed a positive result in order to legally seize our animals. Without evidence of a disease, the USDA could only quarantine our sheep. They could not lay a hand on them unless the secretary of agriculture issued a declaration of emergency, and in order to do that, he needed positive test results.

In lieu of positive results, the USDA knew the best way to manipulate a situation was to create fear in the general public. Detwiler traveled the world, published papers, gave talks about TSEs, and boasted how wonderful the United States' BSE testing program was. After all, she was the USDA's TSE expert and a trusted advisor to the National Cattlemen's Beef Association. Detwiler needed to reassure cattlemen that the USDA was safeguarding their interests, protecting their pockets with the "firewalls" the USDA had in place for preventing BSE from ever occurring in the United States. Everywhere she went, she discussed the Vermont sheep and their threat to the American livestock industry.

When Governor Howard Dean spoke out in our support in November 1999 and said the various government agencies were at cross purposes, that he did *not* want to see the sheep killed, there was *no* public threat, and it was not clear to him who or what was behind the scare tactics, the USDA spin doctors immediately went to work. Dr. Bob Rohwer from the Veterans Affairs Medical Center in Maryland and a close friend of Detwiler's sent a three-page letter to Governor Dean alerting him to the "hazard posed by several Vermont sheep flocks" and beseeched him to have the animals either "destroyed and incinerated or shipped back to their source countries."

"I recognize that these animals represent valuable breeding stock; that their exposure is no fault of their owners; and that such a measure is inherently unjust to those owners. However, to continue to harbor them in the U.S. exposes us to a truly horrible disease of humans and animals and thereby jeopardizes not only the public health but the North American sheep and cattle industries as well. Even if the probability that these sheep are infected is small, the consequences of introducing this disease into North America are so dire that no other course can be justified."

Pretty scary stuff, especially if you are an elected official who's responsible for the safety of your constituents. But the health scare was created by the USDA—not by our sheep. The USDA abused the precautionary principle.

Rohwer's letter gave his credentials and more details on TSEs and their fatal outcomes. He implored Governor Dean, "While the concept of eminent domain runs counter to the Yankee values of Vermont, I can not think of a more appropriate application of the principal [*sic*]. It can not be possibly worth the risk to our public health and agricultural sector to gamble on the unknown exposure history of these animals."

Vermont Commissioner of Agriculture Leon Graves quickly followed with his own letter to Governor Dean. Graves was aware that the governor would be meeting with Mr. Freeman and wanted to beat him at the pass. Again, the letter reviewed the history of our importations and TSE research. Graves said "it was common practice in some countries, including Belgium, to feed meat and bonemeal to sheep . . . and MBM was purchased as a cheap source of protein by numerous countries" (including the United States). He then made a connection between the feeding of MBM and our sheep (because he was unaware, I assume, of the feed records Detwiler had on her desk).

Graves's dire warnings to Governor Dean were, "The repercussions of a positive BSE diagnosis in the U.S. would be *devastating* to our livestock industry. Publicity in itself can be damaging, undermining consumer confidence. We consider not just the contagious aspect of this disease potentially being present in some of these sheep, but the public health significance. Perception is reality regardless of actualities."

To add the final nail to the coffin, Detwiler sent Howard Dean a response she received from Scott Sindelar of the Foreign Agricultural Service when she asked Sindelar for comments on the Korean and Japanese beef markets.

> Officials from the Korean National Veterinary Research and Quarantine Service contacted the Agricultural Affairs Office at the American Embassy in Seoul, Korea, regarding recent articles in the U.S. press that describe USDA's plan to eradicate 365 sheep in Vermont due to concerns over BSE. The officials were apparently seeking some assurances that the United States does not have a BSE problem.
>
> Korea purchases approximately $245 million worth of U.S. beef annually (1994–1998 average) and is the third largest export market for U.S. beef. The Koreans closely monitor sanitary and food issues related to their beef imports and will continue to follow the Vermont sheep issue.
>
> Japanese officials are also monitoring the Vermont sheep case. Japan is the largest export market for U.S. beef and import approximately *$1.4 billion* annually (1994–1998 average)." (My emphasis added. The USDA was planting fear in the minds of our trading partners and then attempting to use this fear as an excuse to destroy our sheep. The USDA was jeopardizing export markets—our sheep were not.)

Detwiler attached a note to her letter to Dean that read:

> Historically, these countries have been extremely conservative regarding BSE. They suspended trade with Canada after the detection of one imported BSE case. The trade was

suspended until the herd of origin was entirely destroyed including those cattle that had been sold from the herd. In addition, Canada had to trace and euthanize the remainder of the UK imports as well as the first generation progeny of the imported dams before trade was resumed. I am providing this information to illustrate how our trading partners monitor events surrounding BSE and how conservative they may be.

What was Dean to do? Or Congressman Bernie Sanders, or Senators Jeffords and Leahy? Risk the multibillion-dollar beef export market for a few flocks of sheep? Senator Leahy decided he would go to Congress for a special appropriation. If he could get Congress to agree to put up a few million dollars to give us and the Freemans, this whole issue could be put to bed and American beef markets would be safe—from a disease that did not exist.

It was Bob Faw of NBC News who called us on October 30, 2000, and told us Congress had appropriated $2.4 million. I was amazed and grateful to Senator Leahy and told Bob this. The money would help toward our loss of business due to the quarantine and Jan Carney's warning against our cheese, and our mounting legal expenses.

"So you're okay giving up the sheep?" Bob asked.

"What? Why would we give up the sheep?" I was confused.

"Because Congress has a stipulation. In order to receive the money, the animals have to be destroyed on or before November 17, 2000."

I was stunned. Why would they do that? Leahy said the money was "to help compensate Vermont sheep farmers who *may have* their herds confiscated by the USDA because of the presence of mad cow disease." He didn't say it was compensation for having our animals killed even if no presence of the disease was detected.

"We won't take the money then," I told Bob.

"Are you sure?" he asked.

"I'm positive."

"They just don't get it, do they," I said to Larry, shaking my head sadly as I hung up the phone. Our fight was not about money, it was about standing up for what was right. We would not take taxpayers' money in exchange for our sheep when we knew there was nothing wrong with our animals. Animals should not be killed without justification, businesses should not be destroyed at the whim of the government, and the USDA should not have the legal right to unfairly harass and seize an American citizen's property.

And what about equal protection? By 2000, more tests had been run on our sheep than on the entire national cattle herd. Cattle that were imported from Europe were still on farms across the United States, even here in Vermont. The animals were under quarantine, but the USDA was not moving in for the kill, the farmers were allowed to continue selling semen from the quarantined bulls and milk from the cows, and the cattle were allowed to live out their natural lives on the farms.

Our sheep were innocent, and yet now there was a bounty on their heads.

Larry put his arm around me and pulled me close, and the tears began to flow. "What do we do now?" I asked, my voice muffled by his sweater.

"We do what we always do: We keep fighting. Who knows what they will offer," he said quietly. He was right. Just because the USDA now had an appropriation didn't mean they would give it all to us. Or any of it, for that matter.

"But we're running out of options. I appreciate Leahy going to all this effort, but why couldn't the sheep go back to Belgium?" I couldn't say anything else.

"We need to talk with the children," Larry said.

That night after dinner we sat around the table and told the children about the $2.4 million.

"Yeah? So what?" Francis asked with an edge in his voice. "They're full of crap."

"You're not thinking of selling out are you?" Heather demanded.

"I don't know," I said, unable to hold her glare. "We seem to be running out of options. The USDA won't let us send them back to Belgium, our choices in court are limited, and our only hope is that Judge Murtha will see through the USDA's lies and rule in our favor."

"You would take a million dollars to let them kill Moe?" Jackie asked with a hard glint in her eyes.

"No! I don't want to give up our sheep for money, but what if the judge rules against us? What if the USDA is able to seize the animals, and then *they* determine what fair market value is?" I could feel the hysteria rising within me.

"Plus all three of you will be going off to college soon, and we need to think about that," Larry added.

"Didn't you say that our outstanding business loans were around $450,000?" Jackie asked, already starting to do the computations.

"That's correct," I said, nodding my head.

"Say the USDA gave us a million dollars, how much would we wind up with?" she asked, her pencil ready, looking much older than her thirteen years.

"Taxes would probably be about 40 or 50 percent," Larry said.

Jackie quickly did a series of calculations. "That means we would receive somewhere between $500,000 and $600,000 before we paid off our loans. Is that worth killing our sheep for?" she asked with tears now in her eyes.

"So, what do you want to do?" Davis asked but probably already knew the answer. We had been through this before when the USDA asked us to submit an offer and we sent a copy of the formal appraisal we had for our sheep—$11,843,345—that was our definition of "fair market value" for our animals.

"I tell you what. I think the USDA should send the sheep back to Belgium *and* pay what they are worth," I said. I was tired of feeling railroaded, forced to play by the USDA's rules.

Davis smiled. We were definitely feisty, a trait he easily identified with. "I thought you would say that," he said.

"And if they don't go back to Belgium and we still have to give up the animals, I want the USDA to test every single one and share all the results with us," Larry said.

"And they should indemnify you against any claims arising from fear of mad cow disease or any TSE," Davis added.

When Davis sent a letter to the USDA's lawyers outlining our requests, the USDA responded, "We have to reject your suggestion about a possible settlement in this case. As I have said before, the USDA will not reconsider sending the sheep back to Belgium. In light of your letter, I can only assume that this aspect of the proposition is critical to you. Since we cannot reach any agreement on that score, I presume it remains useless to meet and discuss the other issues."

. . .

The debate at home continued. The children were adamant we should not surrender the sheep regardless of the amount of money the USDA might offer, I was unsure what to do, and Larry wanted to take care of his family *and* his flock.

Davis told the USDA we would continue discussions, despite their claim it would be "useless." So we waited to hear from the USDA. And waited. On November 15th, the USDA finally responded and requested the following information: "Any previous offers for sheep sales, milk production records, purebred registrations, progeny records, lambing sites, cheese production and sales records, land ownership records, number of ewes by lactation cycles, average milk production per season, uses for milk other than cheese, length of cheese production season, costs of producing cheese per pound, price received for cheese per pound, total investments in buildings and equipment (by item), age of buildings and equipment (by item), other depreciable investments and date, value of land (tax base), and copies of existing and binding contacts with input suppliers. We would welcome a meeting among the relevant parties and/or any submission from you on these matters."

We spent the entire day pulling all the figures together. The USDA met with us on November 16th—twenty-four hours before Congress's imposed deadline to make a settlement.

It was a typical, gray November day. The receptionist at the Federal Building in Burlington escorted Davis, Tom Amidon, Larry, and me to a conference room where again we waited. After twenty-five minutes, the USDA attorney Paul Van de Graaf came to the door and asked Tom and Davis to follow him. Larry and I had to remain in the conference room. Davis looked at us quizzically and shrugged his shoulders. Fifteen minutes later they were back.

"What's going on?" Larry asked.

"They are negotiating in separate rooms. We are in one, the USDA is in another. This is ridiculous," Davis said. No congressional representatives were there, not even someone from Leahy's office. The USDA was trying to create tension. There was nothing collegial about this meeting.

"The USDA still won't name a figure, and they said absolutely no on the sheep going back to Belgium," he continued. "But we had suspected that."

"Then tell them all the sheep have to be tested, we want a portion of brain tissue to have independent testing, and we want to be indemnified against any lawsuits," Larry said.

They left the room and returned twenty minutes later.

Davis said the USDA refused on all points. They would *not* agree to test all the sheep, there would be *no* sharing of tissue samples, and they would *not* indemnify us.

"Forget it then!" Larry exclaimed. "If they won't test the sheep, they can just forget the whole thing," and he stood up to leave. We had wasted enough of our time with the USDA, and today was no different. The USDA had no intention of giving us any part of the appropriation. They were playing games, using money as a weapon. I doubt this is what Congress had intended.

"Hang on," Davis told him. "I need to go back and tell them what you said."

I was relieved. The USDA had made it even easier than I expected. We could go home and tell the children we did not "sell out." Tonight I would sleep well.

Larry and I had our coats on and were ready to walk out the door when Davis and Tom returned ten minutes later. Davis looked more serious than I had ever seen.

"You might want to sit down," he said. I sat, but Larry remained standing beside me.

"The USDA put all the money on the table," Davis said.

"All $2.4 million?" Larry asked.

Davis shook his head. "No. They put the $2.4 million plus $1.7 million they had in their budget for paying you."

My sense of relief was totally gone.

Davis described the scene when he went back to the negotiating room. A different USDA lawyer, one he described as "like a pit bull," was exasperated when we refused to continue negotiating and said the USDA would throw everything in. "You can tell your clients to fight it out among themselves as to who gets how much, but there it is," she said angrily.

We called Mr. Freeman that evening. "I heard about the meeting today," he said. "So what are you going to do?" I was silent. How could I tell the man who was our biggest supporter that we would settle? We had been in this together from the very beginning, and the thought of disappointing him was heartbreaking. But the money would pay off all our loans and help put the children through college. I walked into the bedroom with the phone. "I think we are going to have to settle," I said, my voice cracking.

Mr. Freeman was quiet. I started to cry as I interpreted his silence as displeasure.

"Linda," Mr. Freeman said gently, "I want you to take it all. I want you and Larry to take the $4 million or whatever it is they offered, and I will continue the fight."

"Are you sure?" I asked in disbelief.

"I'm sure," he said kindly.

"You won't be upset with us?" I asked.

He chuckled. "Of course not. You could use the money, and you deserve it for all you've been through."

I walked back into the room. "Mr. Freeman said take it all. Take it all, and he will continue the fight." But there was no feeling of happiness, only a feeling of defeat. Larry hugged me. Our three children glared.

No matter what we said, the children refused to budge. They recognized the corruption of the government and felt that even $4 million was not worth the price of killing the sheep. I had to admire them for their tenacity. Most teenagers would jump at the chance to have that much money (as would most adults), but the principle was more important to them. I didn't sleep that night. The children stayed in their house and didn't even come over to the apartment to kiss us good night.

"We're doing what's right. Aren't we?" I asked Larry as I watched the stars out the window.

"I don't see what choice we have. Your number one priority since the children were little was for them to have a good college education and with this money they will."

I knew he was right, but then I thought of the sheep in the barn. I didn't even get a chance to see them the entire day, and now I was going to "sell out," as Heather said. Kill all our beloved animals in order to put the children through school?

The night seemed never ending, and we called Davis as soon as he was in his office and told him about the conversation with Mr. Freeman. Davis said he would get back to us after talking with the

USDA. The hours dragged on that morning, and around 11:00 the phone rang. It was Davis.

"The USDA said they will only give you $800,000 unless Mr. Freeman surrenders," he said.

"$800,000?! No way. Tell them they can—," Larry yelled.

Davis laughed. "Okay."

Half an hour later he called back. "They've upped it to a million. What about a million dollars?"

"Absolutely not," Larry said.

"That's their final offer, they said. Are you sure?" Davis asked quietly.

"Positive," Larry answered.

The sense of relief was real this time. With great joy I told the children the news when they came home, but it took days before Jackie would speak to me.

20

DETWILER STRIKES AGAIN

We weren't the only ones being attacked by our government. In mid-November I had received a phone call from a woman in Alberta, Canada, who had imported Texel sheep from Denmark in 1994. Six years later the Canadian Food Inspection Agency (CFIA, Canada's equivalent to the USDA) demanded her animals be destroyed. Their reasoning? Denmark had a single case of BSE—in a cow. The woman told me about other shepherds and cattlemen who were also being harassed.

A few days later Don Muroc, a reporter from Vancouver Island, British Columbia, e-mailed me a story about another small family farm under attack by the CFIA. This time it was water buffaloes. Darrel and Anthea Archer imported water buffalo from Denmark with the assistance and approval of the CFIA. Water buffalo milk is used to make the world's best mozzarella, and the Archers' animals would form the basis of a dairy herd for making cheese on their family farm.

Even though water buffalo may bear a slight resemblance to cattle, water buffalo have a different number of chromosomes, cannot interbreed with cattle, and have never had BSE. The Archers informed the CFIA officials of these facts and showed them evidence their animals were never fed any potentially contaminated feed, but their rationale fell on deaf ears. The CFIA was determined to destroy the water buffalo. The Archers resolved to fight back and were going to court in early December. Larry and I decided I would fly to Vancouver to meet them and attend their court hearing in the hope it might have some bearing on our case.

It was early when I walked from my hotel to the Federal Courthouse in downtown Vancouver, British Columbia. Few people were around on this sunny December morning. Once inside the building, I asked for directions to the court hearing. A crowd had already gathered outside the door, and when it opened everyone hurried to find a seat. Feeling particularly self-conscious when I realized I seemed to be the only one showing up alone to the hearing, I quickly surveyed the room and decided it was safest to sit among the reporters, where I might blend in with my notebook. Twenty minutes later, a middle-aged couple accompanied by a young man dressed in a suit approached the area reserved for court participants. It was Darrel and Anthea Archer. Darrel walked quickly, with a slightly stooped gait, and smiled a shy smile toward some of the people in the audience. Anthea had short white hair and rosy cheeks and carried an armful of papers. Her glasses were no mask for the concerned look on her face.

I watched in amazement as Judge J.D. Denis Pelletier and his assistants entered the room dressed in black robes and wigs. This was unlike any American court hearing I had seen. Seated at the defendants' table were the lawyers representing the CFIA. The hearing commenced promptly at 9:30 A.M.

Simon Fathergile, the Archers' young lawyer, appeared to be very nervous. When the judge finally questioned him as to why he was agitated, Simon replied that the matter was so complicated he was worried he would not be able to do the issue justice in such a short time span. The judge told him to take all the time he needed. The hearing would last the entire day.

Simon explained to the court how the Archers, as a measure to save their family farm, imported nineteen purebred water buffalo from Denmark in January 2000. It was the first ever importation of water buffalo into Canada, and the Archers had mortgaged their farm to finance it. Denmark had a case of BSE in February 2000, and six months later the CFIA quarantined the water buffalo, even going so far as restricting the animals to a barn and denying them access to the outdoors to graze.

The Archers waited for the quarantine to be lifted and were shocked when they received a notice on September 1, 2000, to "remove the water buffalo from Canada before the deadline of midnight September 15, 2000, or they shall be forfeited to Her Majesty in right of Canada and may be disposed of as the Minister may direct in accordance with Subsection 18(4) of the Health of Animals Act."

Chills ran up my spine when I heard this.

Simon told the court that if the Archers chose to send the animals back to Denmark they would not be compensated, and if the CFIA killed the animals the Archers would receive "fair market value." "Sounds eerily familiar," I thought. The water buffalo were the Archers' primary means of support, and if they lost their animals they would lose the farm that had been in their family for three generations. So the Archers' hired a lawyer and were able to receive an extension. But on October 5th the entire ordeal was repeated.

A CFIA veterinarian went to the Archers' farm and handed them two letters. One said the original order to remove the water buffalo from Canada was now revoked, and the other said the animals were scheduled to be killed on November 6 and the Archers were to deliver them to Alberta for destruction. Simon argued that the CFIA decision maker never saw the records that demonstrated nothing was wrong with the Archers' animals, and therefore his decision should not be valid.

The similarities between our stories were shocking: government agency approves the importation of animals from Europe, allows the animals to live in the country for a period of time before agents from the agency claim the animals could have eaten contaminated feed and be susceptible to BSE (ignoring the fact that BSE has never occurred in either species), agency encourages animals to be removed from the country but then reneges on their offer and uses legal means to attempt to seize and kill the animals. For the Archers' case it was section 48 (1) (a) of the Health of Animals Act, which said the minister may dispose of an animal that is "suspected of being" contaminated by a disease, and included the withholding of information from the decision maker, all the while claiming they were following the precautionary principle.

"You can't seize animals on the grounds of suspicion. You must have reasonable grounds," Simon continued. "How do you prove a negative?"

This was sounding more and more like our case: theoretical, speculative risk, absence of evidence, and how *do* you prove a negative?

When Don Muroc wrote about the case he said, "So now the politicians have turned this into a witch-hunt. Someone suspects you of being a witch and, unless you can prove you are not a witch, you burn. The same government is irrational about scrapie. During 1997 alone they [CFIA] discovered forty-seven cases of scrapie in sixteen Ontario and Quebec flocks. They destroyed all sheep in ten flocks, but left the rest. On the Prairies, the CFIA found chronic wasting disease in four herds. They destroyed one herd, part of another herd, and left two herds intact. What is their policy?"

Simon told the judge. "Europe is in the throes of the BSE epidemic because of cases in native bovine animals, not imported water buffalo."

"That's a good point for our case, too," I thought, writing more notes in my notebook. I was lost in thought when Simon held up a copy of the American BSE Emergency Response Plan and said, "Canada defers to the United States in making decisions."

After recess it was time for the lawyer representing the CFIA to rebut. "The Crown has the right to be wrong," he said emphatically, "and our decisions are not liable to judicial review."

"Oh boy," I thought, "here we go again."

The CFIA lawyer continued, "Parliament has conferred on us the power to seize these animals, My Lordship."

"Did you speak with the Archers before issuing the order?" the judge asked.

"No, Lord, we did not," the lawyer said. "But as harsh as this order may be financially on the owners, public health and safety is more important. Canada cannot risk having BSE, and the feed records from Denmark are hearsay," he continued.

Quite a few of the Archers' supporters gasped and looked ready to jump out of their seats at that comment. I saw Anthea's shoulders stiffen.

Obliviously, or just plain obnoxiously, the CFIA lawyer went on to say that although the Archers argued that their water buffalo were never even fed grain, "Your Lordship, even though the water buffalo may never have been fed any grain which could potentially contain contaminated feed, if they are exposed to grain, they will eat it."

"What a load of crap!" one of the women sitting a few rows over from me said loudly.

The judge looked her way but chose to ignore her outburst.

Despite the lack of evidence to suggest anything was even remotely wrong with the water buffalo, the CFIA wanted the animals slaughtered to protect Canadian agriculture from a possible outbreak of BSE. The CFIA lawyer reviewed three previous cases of farmers attempting to fight the CFIA. They all lost. Even when a judge found in favor of the farmer, the government appealed and won, "because the judge did not have the jurisdiction," the CFIA lawyer insisted.

He urged the judge to relinquish all control to the CFIA and let them go about their business of protecting Canadian human and animal health. "Legislation grants CFIA uncommon duties and powers to prevent the spread of disease," he said. "And Canada's trading partners will refuse to buy Canadian beef worth billions of dollars if there is a case of BSE found in this country." (The United States is the largest importer of Canadian beef.)

The Archers' hearing ended late in the day, the same way ours had, with the judge declaring he would issue his opinion, "all in good time."

Don Muroc summarized the battle best: "With an arrogance matched only by their ignorance, Canadian politicians appear determined to prove, as have their brethren in Britain, that mad cow disease creates mad politicians."

I spent the next few days at the Archers' amazing farm with its rambling farmhouse and collection of barns and outbuildings. Every morning we gathered for tea and breakfast in the formal dining room and shared our stories. Anthea spent her days writing to elected officials, answering the phone, which rang constantly, and giving interview after interview. Darrel tended to the water buffalo and the rest of the farm animals and resulting chores.

I quickly made friends with the Archers' friends and neighbors. Madia was from Russia and was intelligent and beautiful and spoke with an extremely colorful vocabulary. "She would get on well with Edie," I told Larry on the phone one night as I described meeting everyone. Henry was married to Jill and had a six-year-old son. He was interested in making cheese from the water buffalo milk. Edith was sixty-two-years young and entertained me with stories about her travels and life in Germany. When they discovered it was my thirty-fifth birthday, I was treated like royalty and given wonderful gifts.

In the evenings we either gathered in the Archers' sitting area with a glass of wine or cup of tea or went to a friend's house for dinner. The conversations were reminiscent of the ones Larry and I had at home with our friends. "Why are they doing this? What can we do? Is there any way to stop it?" It was hard not to feel depressed when the answers were the same and a positive resolution felt like it was slipping through our fingers with every day that passed. But what encouraged me was the fact that these people were supporting the Archers and trying to find justice, just as our friends back home were.

One evening Anthea asked me if I had heard of someone from the USDA named Linda Detwiler.

"Of course. Why?"

"The CFIA said they were pressured by her to get rid of our water buffalo."

I practically spilled my drink. "You're kidding!" I gasped.

"No," Anthea said. "I wish I was. Our daughter is going to veterinary school in Ohio, and a few weeks ago she attended a conference where Linda Detwiler was a guest speaker. When the session was opened for questions, our daughter asked if Dr. Detwiler was familiar with the Canadian water buffalo that were under quarantine. Dr. Detwiler replied, 'Yes, but I am seeing to it that they are destroyed.'"

The trip to Vancouver Island reenergized me. Visiting with the Archers and watching their court case gave me hope that we could find a positive resolution with the USDA. I decided to call Bruno Oesch, one of the founders of the Swiss biotech company Prionics, which had developed a highly successful "rapid" test for TSEs. Their test was validated and used by most European governments for their BSE and scrapie testing. Maybe Prionics could test brain tissue from the four sheep the USDA had claimed were positive. I told Bruno about our court battle and sent him Rubenstein's results and asked if Prionics would retest. Bruno said he would and that their test was even validated for sheep.

But when this offer was put forth to the USDA, they not only rejected it, they said there was no remaining brain sample at all. We knew that Rubenstein had "inadvertently disposed of" his samples, but the USDA lawyers told Judge Murtha there were brain samples remaining, just not from the same section that Rubenstein and NVSL had used. Where did they go?

Christmas and then New Year's passed, and still no word from Judge Murtha. Dr. Tom Pringle encouraged me by saying that "every month that passes with the sheep not exhibiting clinical symptoms is in your favor." The sheep were doing very well and were gathered in their barns (ewes and lambs in one, rams in another) for the winter. Our hay supply was abundant, the winter was milder than others but with plenty of snow, and we settled back into our farming routine—with the exception of interviews. This was a part of our life I had come to accept as normal, and by now the entire family was proficient at talking with the media. I didn't realize the impact it had on the children until Heather complained one day that she was stressed because of "homework, farm chores, cross-country skiing, and all the interviews!"

In January we heard that the Archers had won their court case and stopped the CFIA, for the time being, from taking their water buffalo! Our hope was renewed, but short-lived.

On February 6th Judge Murtha found in favor of the USDA and ordered us to "comply with the Secretary's Order of Extraordinary Emergency forthwith." Tears poured down our faces as Larry and I attempted to read his decision.

Judge Murtha felt that the secretary of agriculture had enough information to make his decision and quoted the USDA repeatedly throughout his memorandum of decision. One of the most disturbing errors was the USDA quote, "We recently have been informed that some of the sheep actually originated in the Netherlands and that all of the sheep were imported with no certifications concerning their feeding history."

The USDA knew the animals were from the Netherlands when they approved our importation! All the animals had birth certificates, and the country of birth was listed on them. And there was absolutely no requirement to provide any sort of "feeding certification." Nowhere in the import protocol was feeding even mentioned! We had certified the feeding on our own.

"If the sheep had remained in Europe and had been presented for slaughter, the European authorities probably would have prohibited tissues from these animals from entering the human food supply." The USDA counted on Judge Murtha's ignorance for this statement. The USDA was referring to the fact that Europe removed specified risk materials (SRM) from all ruminants. SRM were tissues that had the highest possibility of being infected with the BSE agent, such as the brain, spinal cord, eyes, tonsils, small intestines, and so forth. Europe was proactive about protecting human health from any possible risk from the BSE agent and prohibited these materials from entering the human food chain in 1989 and extended the ban to include mammalian and pet food chains in 1990. SRM from sheep and goats were banned in Europe in 2000. Meanwhile, as of 2006, the USDA has still *not* banned these materials from the human, mammalian, and/or pet food chains.

Judge Murtha also referred to the Geographical BSE Risk Assessment of Belgium and the fact that "until 1994 the BSE/Cattle system was extremely unstable because rendering practices would have allowed BSE infectivity to survive and the feeding of MBM to cattle would have amplified infectivity." This situation improved slightly with the introduction of the feed ban in 1994. Yet Judge Murtha was unaware that the United States did not even *have* a feed ban in place until 1997, and an investigation in 2001 revealed that more than 70 percent of all feed manufacturers were not in compliance with the regulations.

Regardless of the feed records and all the negative test results, Judge Murtha felt absence of evidence was not evidence of absence and ruled that Secretary Glickman was correct to order our animals destroyed.

We immediately applied for an appeal and a stay to stop the USDA from taking the sheep. By law, the USDA had to give us ten (business) days to file an appeal. The USDA notified us they intended to seize our sheep on February 20, 2001.

21

PROTESTING SNOWMEN

Adrenaline kicked in once again. The entire family and community were on overdrive—there had to be some way to stop the USDA. Together we tried every possible tactic: Save our Sheep (SOS) meetings were held with increasing frequency; SOS bumper stickers were seen on cars up and down the East Coast. (My sister Mary called me from Washington, D.C., one morning to say the car ahead of her had Virginia plates and an SOS bumper sticker.) Rural Vermont was a farm advocacy group, and staff member Alexis Lathem wrote many amazing, articulate articles in our support and attended most of the SOS meetings, despite living an hour away. Edie Connellee organized fundraisers at American Flatbread to help pay our legal fees; many, many people including Edie, Barry Simpson, Jim Leyton, and Ruth Joslin wrote letter after letter to newspapers and to our Vermont elected officials urging them to stop the USDA; people from across the country called in support; Jim McRae, former president of the Vermont Sheep Breeder's Association and fellow sheep farmer, established an SOS website with the help of his friend Don Green; John Barkhausen, Marc and Jacki Harmon, Roy Morrison, and Bruce Fowler formed FACT (Farmers and Citizens for Truth), an organization that spoke not only on our behalf but looked at the larger issues behind USDA corruption.

On February 16, Judge Murtha turned down our second request for a stay but ordered the USDA not to seize the sheep until February 22 to allow us time to appeal to the second circuit court. The USDA sent Davis a letter saying they would take Mr. Freeman's sheep on February 24th and our sheep on the 26th.

On February 22, SOS and FACT hosted a rally on the Vermont statehouse lawn. It was a beautiful, sunny winter's day, and a crowd of more than seventy people braved the 10-degree temperature to show their support. The media coverage was excellent and Jim McRae, Roger Hussey, John Barkhausen, Barry Simpson, Kinny Connell (our local state legislative representative), two local farmers (Anne Miller and Joe Klein), Maine sheep farmer Pam McBrayne, and Vermont beekeeper Ross Conrad gave rousing speeches from a podium on the steps leading to the statehouse.

"Continued insistence on the destruction of the Faillace and Freeman sheep has gone beyond preposterous," Barry Simpson told the gathering. "The information relied upon by former Agriculture Secretary Glickman in issuing his extraordinary emergency order in July was obsolete, invalid, and heavily biased at the time of his decision." Cheers of agreement came from the crowd.

"Furthermore his department may have mishandled and certainly concealed a large body of evidence that if viewed objectively at the time would have precluded issuance of the order, or certainly would have enabled Judge Murtha to void it at the first opportunity.

"The USDA's misdirection on the mad cow issue opens up a brilliant opportunity for Vermont agriculture, however. A public relations bonanza would follow the revelation that not only are the world-famous sheep flocks entirely free of mad cow disease or any variant but that Vermont is beginning to test all cattle old enough to slaughter for mad cow disease as a demonstration that our state is entirely free of BSE infection." And with that the crowd roared.

After the speeches, everyone marched from the statehouse to the federal building to the department of agriculture building. Accompanying the group were Jacki Harmon and two friends in borrowed sheep costumes from Bread and Puppet Theater. It was a peaceful but invigorating rally. That evening the Vermont Historical Society called and asked for some of the protest signs. They knew history was in the making.

A little over a week later Larry and I were back at the Vermont statehouse. In 2000 Bruce Hyde, our local representative at the time, had submitted a resolution "urging the USDA/APHIS and the Vermont Department of Agriculture to expedite the removal of the quarantine imposed on two Vermont sheep flocks located in Warren and Greensboro." It had passed the Vermont House but remained suspended by the Senate Agricultural Committee.

Exactly one year after Bruce's resolution, Kinny Connell introduced another resolution to the Vermont legislature that called for the

"USDA to rescind its order to destroy sheep flocks in Warren and Greensboro and to negotiate a scientifically based and satisfactory settlement with the flocks' owners."

Larry and I went before the Ag Committee at the statehouse, and Larry gave the following speech:

First a quote from Ralph Waldo Emerson, which Linda and I find helpful: "Write it on your heart that every day is the best day of the year." We are here today to propose reason and science. When a scientific issue is treated in a political way, it is bad for everyone. When hysteria prevails over clear thinking, it is even worse.

Representative Kinny Connell of Warren has introduced Vermont House Resolution 12. H.R. 12 calls for reason, it asks former Secretary Glickman's order for the sheep to be seized and destroyed to be rescinded, and it calls for science; for a real solution to be formulated using clear scientific thinking.

This is certainly not a new idea. In fact it is what we proposed two and a half years ago when the USDA first met with us at our farm and expressed their concerns. We suggested setting up a testing program. Over the past few years, we and Mr. Freeman have attempted to put forward ideas and compromises that could verify the fact that the sheep are healthy, assure consumer confidence, and allow family farms to once again prosper.

What is new today is the ever increasing support we are receiving from friends, farmers, and the American public to have this issue resolved in a humane and scientific manner.

Before I go any further I want to emphasize a really important point. Linda and I are as concerned about mad cow disease as anyone else, perhaps more so. We lived in England at the height of the mad cow scare and witnessed the effect it has on people's confidence in the food system as well as the devastating effects it had and continues to have on the farming community.

During our time in England, working at the University of Nottingham School of Agriculture, Linda and I were not only exposed to the issues surrounding mad cow disease but were active participants, working directly with Professor Eric Lamming, a scientist commissioned by both the British Parliament and EU Commission to give advice regarding the role of cattle feeding practices.

Our work in England taught us the seriousness of the issues surrounding mad cow disease. That knowledge was very much in mind when we began working with the USDA late in 1993 on the prospect of importing European sheep to help form the foundation of a new industry in Vermont.

The most important thing the study of BSE has taught us is BSE is not a disease of sheep. That theoretical possibility was first addressed in 1993, but after eight years of surveillance and experiments with European sheep, not a single case has been found, not one.

We used our knowledge and experience and went above and beyond the USDA's rules of safety for importing the sheep. The USDA had no provision in their protocol for what our imported sheep should have or should not have been fed, but we knew it was important. Specifically, Linda and I made it a requirement that the flocks of origin were under official surveillance for scrapie and that meat and bonemeal had never been fed to the imported sheep or their ancestors—here are the certified feed records to prove it.

Prior to our import of the sheep, there were no sheep flocks in Vermont being officially monitored for scrapie. As a requirement of the import, and more importantly because of our strong belief in the control and eradication of scrapie from the sheep population, we helped begin the scrapie-monitoring program in Vermont—really quite an irony considering the predicament we are in. An unfortunate consequence of this situation is that we have had many sheep farmers, not only in Vermont, but around the country, tell us that they are fearful of joining the USDA's program. However, by working with the USDA to implement a scientifically based testing program, confidence in the scrapie surveillance system can be restored.

Another important conclusion about BSE, no matter what theory you follow as its cause, is that forced cannibalistic practices in our livestock must be stopped. The United States must follow Europe's example and stop the feeding of all mammalian proteins to ruminants, which are natural vegetarians.

We would like to meet with [USDA] Ag Secretary Ann Veneman to further ideas outlined in House Resolution 12. She has the unique opportunity to undo many missteps that have been made and to solve this issue in a way that upholds scientific principles, truly protects consumers, and, last but not least, protects farmers.

Finally, this battle is no longer about "saving our sheep." It's about much, much more. Family farmers should not have to contend with government personnel that do not follow the agencies' own rules, trespass on private property, mishandle and destroy evidence, and even withhold and misrepresent evidence from the secretary. The USDA is supposed to stand for science as it relates to agriculture, and the actions of the individuals handling this case have tainted the USDA's reputation. Family farmers and consumers should expect the USDA to protect them, and the actions of these USDA officials have protected neither.

I am pleased to report that just over the weekend we had several hundred people contact us in support of House Resolution 12, and we ask the Vermont House to pass this resolution as soon as possible. Thank you for the opportunity to speak.

"Mommy, Daddy, come quick!" Jackie yelled, trying to catch her breath as she ran up the stairs to our apartment.

"What is it, Honey?" I asked, as I stood at the sink washing dishes.

"Quatrini and Christine had lambs. Four of them!"

"What?!" Larry exclaimed.

"Lambs?" I asked, my mind unable to register the thought.

"Four of them," Jackie repeated excitedly.

I quickly dried my hands as Larry grabbed our coats and gloves and we hurried to the barn. Sure enough both girls had a set of twins beside them. Quatrini's were nursing while Christine's were curled peacefully beside her.

"What the heck?" Larry said. Bewildered, we stood together

looking at the lambs as the scent of green hay filled the barn. Larry climbed into the paddock and picked up Christine's lambs. "They're ewe lambs," he said, as he handed one to me.

"How did this happen?" I asked, savoring the feel of the soft lamb in my arms.

"Wasn't a Friesian ram who had fun, that's for sure," he said. Quatrini and Christine were both East Friesian ewes, and the lambs had the distinct "blocky" heads and black noses of a Beltex.

"It must have been when we put all the sheep near each other and Barry Simpson loaned us the light," Jackie piped up.

I made a quick calculation. That was five months ago. Jackie was right.

"Uh oh," I said.

"What's wrong?" Larry asked.

"The USDA is going to be angry. Remember the USDA lawyer sent a letter back in the fall asking us not to breed the sheep, and we agreed? Now they are going to accuse us of breeding anyway."

"Well they should have stayed the hell off our property, then this would have never happened," Larry replied sharply. "Besides, they know we never breed the sheep that early. We always breed around the end of November. Who would want to be lambing in the dead of Vermont winter?"

I nodded and scrutinized the other ewes. Who else was pregnant? Were any of the Beltex ewes pregnant? They had their full winter fleeces and were rotund to begin with, so it was hard to tell. The lamb Larry was holding bleated, anxious to return to her mother. I handed him the other lamb, and he set them down gently beside Christine. I named the lambs "Prosperity" and "Hope."

The next morning, Francis spotted the green SUV with Connecticut plates going up Roxbury Mountain Road. James Finn was driving. The same man who had told us, "You won't see me around here again," when he handed us the seize and destroy order. Yet Francis would identify him on a regular basis, always in the same vehicle—a green SUV.

That night at the SOS meeting, supporters proposed ways to respond to the USDA. Jim McRae thought civil disobedience might help: "The only way to attack this problem is the way the French do. Civil disobedience is the way to go; if it comes to that, I'll be in front of the truck." Radkin, another supporter, agreed. "This is a war that's come right home. If we can't win the battles in our own communities, then we will never win in Washington," he said.

Larry and I hoped such actions would not be necessary, and we wouldn't encourage it, but we also didn't want to discourage people's ideas as long as no one got hurt. When Francis told them about Finn, Radkin suggested getting a motion sensor for the barn. I just prayed nothing would happen to our newborn lambs.

There was a large snowstorm the night before town meeting that year, and Vermonters had to work hard to dig out their cars or use their snowshoes or cross-country skis to attend the New England tradition that allowed individuals to be heard within their community. The Mad River Valley had strong participation in their town meetings, and, despite the blizzard-like conditions, this year was no different.

Larry and I were greeted warmly when we went to the Warren Elementary school and, once the meeting started, I temporarily forgot about our battle and was lulled into conversations and votes about town budgets, elected officials, and discussions on Warren's new wastewater system. My reverie of everydayness was destroyed when Bruce rushed in.

"They're coming!" he said loudly enough for a few rows of people around us to hear. "The USDA is coming!"

Larry and I immediately jumped out of our metal seats. "What do you mean, Bruce?" Larry asked, running over to him.

"I just heard on the radio that the second circuit court refused to issue a stay. So that means the USDA can take the sheep."

"So how do you know they are coming?" Larry asked as he grabbed his hat and gloves.

"Because they can."

Back at home Larry and I called Davis. He was out of his office, taking testimony for another case. It was late in the afternoon when we finally received Davis's call. "I have some good news and bad news," he told Larry. "The bad news is the motion for a stay was denied. The federal appeals court said they did not have the legal right to stop the USDA from seizing the sheep, but the good news is they have agreed to hear your case and have expedited your hearing.

Instead of the usual eight or nine months wait, they have us already scheduled for April 10. Exactly five weeks from today."

Larry thanked Davis and hung up the phone. The gathered crowd in our apartment was silent as he explained Davis's call. No one knew what the USDA would do under the circumstances.

We found out two days later when the USDA lawyers sent Davis a letter. They would ship the sheep to Ames, Iowa, and they would seize them under these conditions:

- No protestors or media allowed on the property
- Three-foot-wide path shoveled between both barns and the road
- All animals gathered in one area
- Remove the llama from the barn
- Give the location of any sheep semen

The USDA refused to say when they would kill the sheep, just that it would take place sometime before April 10.

It was my turn to get irate. How could the USDA touch the sheep before our court hearing? And if they did show up to take the sheep, there was no way I would stop the media or protestors from witnessing the USDA's crime. If the USDA wanted a three-foot-wide path they could come shovel the tremendous amount of snow that had accumulated themselves. (There was so much snow that when it shed from the roof of our house it piled up to the eaves. Walking into the house was like entering an igloo—no sunlight from any windows because mini mountains of snow covered them.)

The animals were already in the appropriate barns, and the llamas were there to guard the sheep. Freddie the llama would be too distressed to be removed from his flock. He would stay.

And we had requests of our own. If the USDA took the sheep we wanted to have tissues to do our own independent testing, and we made arrangements for Dr. Karen Anderson to travel to Iowa and assist in tissue collections. The USDA would have to make special arrangements for the lambs. We now had more than twenty lambs frolicking in the barn, including Kanga.

Francis's friend Brett Seymour was with me in the barn the day in early March that Kanga and her brother were born. Brett had never seen a sheep give birth and watched in silence and fascination. Kanga was born first, and her brother quickly followed. But Kanga's mother, Eternity, was interested only in her ram lamb, the larger of the two.

"Look at the way she's walking," Brett said quietly as Kanga took her first steps. She would wobble along and then give a little hop. "She looks like a kangaroo. You should name her Kanga," he suggested.

Over the next few days it became apparent that Kanga would be our second bottle lamb, so we brought her home to stay in our apartment. The apartment was the center of activity with people coming throughout the day, and Kanga thrived on all the attention. When Larry and I would go to bed, she would stand below our bed, bleating. We learned to shut the door to our bedroom if we wanted to sleep, but we could still hear her racing around the couches every night before finally settling to sleep on a blanket, curled up like a puppy.

Davis sent a response to the USDA's letter and told them our requests. He said we would cooperate with the USDA and that protestors would not be on the property, but the media would be, and he warned that by refusing the protestors access to the property they would be in the road. Davis ended his letter, "I withhold comment on USDA's refusal to disclose the time of seizure until it enjoys the cover of darkness, maximizing the anguish inflicted on the family."

In the meantime, efforts to save the sheep continued. A large group of SOS supporters went to Jeffords', Sanders', and Leahy's offices. An aide at Leahy's office listened to their concerns and said he would "see what could be done." Senator Jeffords' staff replied, "The senator has tried all he can do. Sorry." At Congressman Sanders' office, the SOS supporters and the accompanying media cameramen had the door literally slammed in their faces when they attempted to talk to an aide. This became a top news story as Sanders was the self-proclaimed "friend of the family farmer."

All was made clear a week later with a "Joint Comment by the Vermont Congressional Delegation in Support of USDA Acquisition of the Imported Sheep in Vermont." According to the document, all three of Vermont's Congressional Delegation agreed with the USDA's decision: "Any further delay could endanger public health and put at risk the entire New England dairy and sheep industries. The potential harm could easily go beyond our agricultural industries to also impair travel and tourism and the reputation of the Vermont brand that is so important to our economic future.

"The Faillaces put heart and soul into this project to promote agricultural diversity and value-added agriculture in Vermont, and we commend and admire their vision. Our hearts go out to these families for the hardship that resolving this will entail and for the emotional investment they have made in these flocks. At some later date, if they are still interested in pursuing this dream, Vermonters should come together to help them get a fresh start. We wish there was a sound alternative to the removal of these flocks, but there is not."

Leahy's website also listed the following: "As an example, the Delegation noted, Dr. Robert Rohwer, director of the Molecular Neurovirology Laboratory of the Medical Research Service, said, in a letter to Governor Dean, that the Vermont sheep pose a 'threat of truly horrible disease of humans and animals and thereby jeopardizes not only the public health but the North American sheep and cattle industries as well. Even if the probability that these sheep are infected is small, the consequences of introducing this disease into North America are so dire that no other course can be justified.'"

Yet when USDA spokesperson Ed Curlett was asked by the Associated Press on March 20, 2001, about the risk of shipping the sheep halfway across the country to Iowa, he responded, "The rest of the sheep [besides Rubenstein's four positives] are considered exposed but aren't sick. And the possibility of them spreading disease is as close to zero as you can get."

The phone rang early on the morning of March 21, 2001. "They're taking Mr. Freeman's sheep," Davis said quietly. I felt a lump in my throat, and the tears, which these days were all too ready, started to fall.

"Are you sure?" I asked, hoping there was some mistake.

"I'm sure," he said.

Larry and I turned on the television and watched as armed federal agents and USDA officials gathered all of Mr. Freeman's 238 sheep onto two tractor trailers. Soon the children were up and assembled in our apartment. We stood in silence until Heather spoke.

"So we're next," she said.

Jackie didn't wait to hear anything else and quickly ran down the stairs. I watched out the window as she headed to her beloved sheep in the barn.

Francis slammed a cereal bowl on the counter. "What a load of shit! How can they get away with this?" I moved toward him in an attempt to hug him, but he angrily pushed me away.

"This is supposed to be a great country, but all the stuff we are taught at school is crap. It's all a pack of lies!" Pélé instinctively moved as close as possible to Francis. "We don't have any rights!" Francis continued, automatically petting Pélé, unaware of his hands, his mind trying to fathom reality versus what he had learned.

Then he turned on me. "I don't get it. You're the one who says 'think positive' and 'you can create your own reality.' Well what sort of reality is this?" His eyes blazed in anger.

I had no answer. What could I say? Tears streamed down my face, and I turned away.

"Leave your mother alone!" Larry yelled. "It's not her fault. We all tried."

Francis stormed out of the apartment, leaving Heather, Larry, and me standing in silence.

"Should I activate the phone trees?" Heather finally asked quietly.

"Thanks, Honey," Larry said as he filled the kettle with water for tea.

I sat down at the computer and sent an e-mail to our SOS list and told them about the USDA taking Mr. Freeman's sheep, and we assumed ours would be next. "Our love and gratitude goes to all of you for your prayers, support, and efforts. We will notify everyone as soon as we hear from the USDA. If you wish to witness, please come with your cameras, videos, and blessings for our sheep."

We asked Jim McRae to give his own appraisal of our sheep so we would have evidence of their condition at the time of seizure. That afternoon I videotaped the entire procedure. It was hard watching him assess every animal. Jim attempted to make things lighter and joked about the "pigs with wool" and the fact that the crossbred Beltex lambs could have been linemen for the New York Giants. I wondered how many days more we would have with our precious animals.

Davis sent the following letter to the USDA lawyers that day: "Since the Freeman sheep will be in Iowa shortly, we request that you test appropriate tissue from the parents and siblings of animal numbers

3711, 3714, 3715, and 3706 (the Rubenstein positives). If these animals prove to be free of any TSE there would be no reason to proceed with the seizure and destruction of the Faillace flock."

Davis did not receive a response, and on March 27, 2001, forwarded the letter to Secretary of Agriculture Ann Veneman. She never replied either.

Detwiler was in her glory. Never had she given so many interviews, and, because she was the government's "expert on TSEs," no one dared to question her. She was the authority. She could say whatever she wanted. "USDA had no choice but to take decisive action. We needed to take those sheep," she said emphatically.

She reassured consumers that the sheep did not pose any danger to the food industry. "The farmers sold some cheese and milk products as well as meat from the sheep prior to a 1998 quarantine on the herds. The U.S. Food and Drug Administration and the Centers for Disease Control and Prevention traced the food products, and its investigations found no infection of BSE," she said. (So why was the warning against our cheese *still* on the Vermont Department of Health's website?)

The next morning Larry and Detwiler were interviewed remotely by Diane Sawyer for *Good Morning America*. The children were in the barn, and reporter Chris Noble summarized the scene best:

> A heavy snowfall blanketed the hills and trees in this central Vermont village of 1,200 as Francis Faillace, sixteen, and his sister Jackie, fourteen, broke up bales of hay for the flock. The children were mostly silent as they did chores under the gazes of half a dozen reporters and photographers.
>
> "I don't think there's much we can do to stop it," said Francis, his voice mingling with the gentle bleating. His quiet resignation barely masking his anger over what he saw as heavy-handed government tactics.
>
> "They've lied, they've cheated, they've broken so many laws," he said.
>
> Heather Faillace, fifteen, was surprised by the attention and the magnitude of the forces bearing down on her family. "This is the biggest thing I've ever been through," she said. "We're in a fight with the government."

The *Good Morning America* camera crew quickly and quietly set up their equipment while the children finished their chores, and soon a spotlight in the barn was on Larry as he stood, dressed in his Carhartts and shepherd's hat. He was interviewed by Diane Sawyer for about three minutes, and then Diane had Detwiler on from Montpelier.

Diane Sawyer questioned Detwiler, "Let me ask you point-blank. Are you saying that these tests show there is a form of mad cow disease in this country in these sheep?"

"Yes, we are," Detwiler replied. "We're saying that we stand behind the test results."

That afternoon Detwiler continued her smear campaign by hosting a "press conference" in Montpelier. It was not a true press conference, because Detwiler had her henchmen stationed at the door and would only allow certain members of the press to enter the room. Some of the local reporters that were sympathetic to our view were refused entry. John Dillon was allowed access and asked Detwiler about her comment that the sheep had a form of BSE.

"The USDA has never said this is BSE," she replied, directly contradicting what she had said that morning and then temporarily denying the phrase the USDA was using to cause panic in the American public. That was why Korea, Japan, and people around the world were watching the situation. Everyone wanted to know if the United States had BSE.

When asked when the USDA would seize our sheep, Detwiler replied, "It will be a matter of days or weeks. We're not announcing it publicly."

But it wasn't days or weeks. Wayne Zeilenga called me at 6:25 P.M. that very evening. "The USDA will be coming tomorrow," he said. When I asked for the exact time, he replied he was unable to answer any questions about when or how. He said he wanted to call because it was more personal, and he reminded me of when he brought the quarantine papers and gave them to us in person, not through the mail. I responded by telling him the USDA was destroying the American livestock industry.

The next few hours were a blur as my sisters and friends activated the phone tree and contacted all the reporters and supporters. The children disappeared for a few hours, and I let them be. I figured they needed some quiet time in the midst of all the chaos.

Jacki Harmon and Edie organized a candlelight vigil, and a group of about thirty people braved the snowstorm to stand with candles in a circle outside the barn saying our prayers and good-byes to the sheep. The light from the candles revealed what the children had been up to the past few hours: The rays illuminated six or seven snowmen, one in the shape of the Statue of Liberty.

Each one had a protest sign around its neck.

22

THE SEIZURE

The snow was steadily accumulating. With three feet on the ground and more falling, travel became difficult. But not difficult enough for twenty-seven armed federal agents and thirteen USDA officials to invade our farm that morning.

Larry and I hadn't slept much the night before. Jordan Silverman, a photographer who had closely followed our story, was sleeping at our house, and we warned him of Kanga's nighttime routine. We heard him chuckle as he turned out the light and Kanga proceeded to run circles around the couch where he lay, eventually choosing to cuddle beside him.

I lay in bed watching the snow float past my window, illuminated by the back deck light, and thought of everything that happened over the past eight years: all the planning, traveling to Europe to select the farms and then the individual sheep, and later importing the sheep to our hillside farm in Vermont. Images flashed through my mind of the Beltex ewes waddling through the snow, following the llama to a new bale of hay, of Pélé playing with his "girlfriend" Sweet Pea, of Benedict walking through the woods with Larry and me, Heather taking comfort in milking her sheep, Jackie getting national recognition for her cheesemaking, and Francis managing the pastures so well they looked like a well-manicured lawn. And then I thought of Moe.

As tears silently streamed down my face, I tried not to make any noise on the slight chance that Larry was sleeping. I remembered Moe playing with the rabbit and the guinea pig, running over to greet me when I arrived home, chasing me as I ran to the store, and pawing at my leg so I would pick him up and hold him. I thought of Moe and our futile attempts to reacquaint him with the other sheep, which always ended with him running back to me, not wanting to be away from my side. I couldn't contain myself any longer and started sobbing.

Larry tightened his arm around me. I turned to bury my face in his chest and noticed that his face was wet, too.

At five in the morning we received a phone call from a supporter in Middlesex, Vermont, alerting us the USDA was on their way, and "there were quite a few vehicles in the convoy."

As the children headed to the barn in the burgeoning light to feed the animals, Larry and I spoke with some of the photographers and reporters who were already at our home. Around 5:30, as we walked to the barn, Jackie came running down the road crying, "Daddy! Daddy! They won't let me feed the sheep! They keep telling us we have to get out of the barn."

"What do you mean they won't let you feed the animals?" Larry asked.

We picked up our pace. When we came to the top of the knoll in the road, I stopped for a split second, unable to believe my eyes. Supporters had covered the outside of the barn with signs of protest. Birch trees were painted with "USDA Lies," and the road was spray-painted with the words "Don't let politics kill the sheep" and a large circle with "USDA" in the middle and a slash through it. Federal agents were everywhere, swarming the property in their black bullet-proof jackets with the words "Federal Agent" blazed on their backs, setting up tape around the barn, treating our farm like a crime scene, while I knew it was they who were committing the crime.

"This is like something out of a movie!" I whispered in disbelief to Larry.

Larry stormed over to one of the agents and asked who was in charge. The agent pointed to another man, of similar height and hair color, wearing an identical jacket, talking on a cell phone.

"What the hell do you think you are doing?" Larry demanded.

"Excuse me?" the man replied arrogantly.

"My children are allowed to feed the animals."

"I'm sorry, but *we* have a job to do."

Larry's face turned red, a vein bulged on his forehead, his eyes flashed with anger. The children and I recognized the look. It meant it was time to back off, that Larry had reached his limit. "I don't care what the fuck you think you have to do, my children are going to feed the animals and you are not going to stop them!"

The agent took one look at Larry's face and relented. He, too, could identify someone at their breaking point.

Another agent was in the process of telling Francis to move Freddie, the llama.

"Why?" Francis asked incredulously.

"He's dangerous. The USDA warned us to keep our distance."

Francis laughed, not a kind laugh. "Dangerous? You have got to be kidding!" And he walked away, leaving Freddie to lean down and sniff the nervous federal agent's hair.

Jackie and Heather were now spreading hay among the pens, refilling water buckets, checking on all the sheep. They were doing the same chores, just as they had over the past four years—only this time would be their last. When they thought no one was looking, they let themselves cry. But I saw them, and felt helpless.

"Mom! Dad wants you," Francis yelled from the path.

The road was now jammed with cars, friends and supporters, reporters and their camera crews, and the state police. Larry was standing next to a police car as I walked over.

"Ms. Faillace?" the officer asked.

"Yes?"

"I'm Officer Dimmick from the Vermont State Police. I want you to know that we are here for *you*. To protect you," he said looking toward the federal agents now stationing themselves around the perimeter of the area.

"I appreciate that, very much," I said, extending my hand. This is why we live in Vermont, a place where people take care of each other. "But we should not have to protect ourselves from our own government," I thought, watching as the media scrambled to find the best position among the snowmen. The federal agents had cordoned them off like animals, limiting them to a very small enclosure.

I went back into the barn to find Jackie arguing with a female federal agent.

"What's going on here?" I asked, looking at Jackie's tear-stained face.

"She's getting in everyone's face and taking their pictures. I want her out of here!"

I was stunned. Jackie rarely raised her voice at anyone, let alone flung insults.

But Jackie was right. The woman had a digital camera and a video camera suspended from straps around her neck and was intrusively taking everyone's photos. And for what reason?

"Look, I'm just doing my job," the woman said, as she raised her arms in exasperation.

That's when Jackie spotted the woman's sidearm. Now she was really angry.

"What do you need *that* for?" Jackie asked through clenched teeth, pointing at the holster.

The woman quickly lowered her arms and tried to pull her jacket down to cover the weapon.

"There is no reason you should have a gun!" Jackie insisted.

"This is . . . for . . . protection," the woman stammered.

"For protection against what?" Jackie snorted, her eyes shining.

Jackie and I looked at the other federal agents in the barn and saw they too were armed. What exactly were they expecting?

"Jackie, why don't you help me, the USDA is here," I said, gently pulling on her arm. Jackie turned to look at me, and I saw the softening in her face as tears welled in her eyes. "Where's Hanna?" she asked. Harwood High School was closed for the day due to the storm, and students had been encouraged to try to get to our farm.

"She'll be here any minute. Look, I'll handle the USDA. Why don't you go check by the entrance. The road is totally blocked with all the vehicles, and Hanna probably had to walk up from the four corners."

A group of USDA officials dressed in brand-new Carhartt suits entered the barn. I was surprised to see they had the appropriate attire for the weather. Then I looked at their feet—loafers.

Bill Smith strolled toward me with his hand outstretched. I shook it but made no attempt at small talk. He introduced me to his cohorts and said, "They will be traveling with the animals all the way to Iowa, so if you have any special instructions, please let them know." A large eighteen-wheeler with a red cab and double-decker trailer pulled perpendicular to the entranceway. "All the animals are going on that?" I asked pointing at the vehicle.

"Yes, that's correct," Bill replied.

"You are going to separate them, aren't you?" I asked.

"If you think we have to."

"Of course you do!" I exclaimed with frustration. "You need pens for the rams, another pen for the ewes, and a pen for the ewes with lambs. Plus someone has to take care of Kanga. She needs to be fed every four hours."

Right then Jackie walked into the barn with Hanna. "Jackie, could you make up the milk replacer for Kanga?" I asked. I grasped at anything to possibly keep her mind off what was happening.

I led the USDA officials through the solar barn and described each paddock, each group of sheep, and who could share a pen together. Silently I tried convincing myself that something was going to happen to make this nightmare stop. Senator Leahy was going to show up and say that Congress was investigating the entire issue. We were only seventeen days away from our second circuit court hearing, maybe something would come through on the legal side. What about the letter we sent to Ag Secretary Veneman requesting a reprieve until all of Mr. Freeman's animals were tested? But nothing happened. I had to keep going, explaining the various groups of sheep. I needed to make sure all the animals were given the best treatment possible, right until the end.

And then we came to Moe's pen.

Moe was resting in the back of the paddock, a little removed from the others, still convinced he was a dog, not a sheep. When he heard my voice he jumped up and ran to me, bleating. "This is Moe," was all I could say to Bill and his entourage. Then the tears started.

"Let's go see how the others are doing with the lanes," Bill said, ushering the group out of the barn.

I leaned over the pallet to pick up Moe. He now weighed in at close to seventy pounds, but I was determined. I hoisted him over the enclosure and sat down on a bale of hay. As he nuzzled me I inhaled, wanting to savor every smell and the feel of him in my arms. I was angry and hated the feelings of helplessness running through me. Tears ran down my face as I buried it in his wool, trying to block out the chaos around me.

How could this be happening? How could they kill my Moe?

There was a commotion outside the barn. Realizing this would be the last time I could hold Moe, I tried to ignore the people calling my name, but then I heard my father's voice.

As gently as possible, I put Moe back in his paddock and went to see what was happening. The federal agents and USDA officials had set up yellow police barrier tape, making a lane from the solar barn to the livestock trailer. The federal agents would not allow anyone to cross the line. My parents and two of my sisters and their families were on the other side. I explained to one of the agents that this was my family and insisted they be allowed in the barn.

Sitting so quietly in the corner of the barn that I had not noticed him was Brad. Over the past few years, Brad would arrive unannounced at the farm and spend time alone with the sheep, content just to be in their presence. Now he held Kanga on his lap. There were no tears, just a tremendous look of pain on his face. I went over and put my hand on his shoulder. He looked down at Kanga and held her a little closer.

More than sixty protestors were now gathered in the road, shouting into megaphones and carrying large signs. Everyone was bundled up against the cold, and the snow was still falling at a rapid pace.

"What the hell do you think you're doing?" I heard Larry yell.

I ran outside and saw a front-end loader forcing its way onto our land, attempting to plow a path but destroying the field underneath, the field that we had worked so hard to carefully graze and restore.

"Get that thing out of here!" Larry yelled.

The driver of the machine looked at Larry and then back at the agent who was giving him instructions. The agent nodded for him to keep going.

The front-end loader attempted to muddle forward. Instead of white snow, dark soil was now flying up from the wheels. "What the hell?!" Larry yelled again, running toward the machine.

He turned to the agent. "You tell him to stop right now, or there will be real trouble," he said threateningly.

"We need to make a path for getting the animals," the agent replied.

"You don't know anything about animals," Larry snarled.

By this point the USDA officials noticed the commotion.

"What's going on here?" Bill Smith asked.

"They're destroying our field," Larry answered, pointing at the machine.

"We need to make a path to bring up the animals from the lower field," the agent said to Bill.

"No you don't," Larry said.

"How would you suggest we do it?" Bill asked Larry quietly.

There was already a narrow path from the solar barn to the rams' shed through the snow, packed down by the daily feeding and watering of the animals. It would be easier to lead the animals one by one.

"If you take a bucket with some grain, the rams will follow you any-where," Larry explained. I knew what he said was true. The animals were so tame, they would trust anyone. At that moment I wished we had chosen to import Scottish Blackface or Bleu du Maine—wild, skittish sheep that would scatter as far as possible.

Protestors saw what was happening and offered to surround the vehicle. The driver had a look of panic on his face and feebly exclaimed, "I'm a Vermonter!"

"Then what the hell are you helping them for?" was the response he received.

"Tell him to turn it around. We won't need it," Bill told the agent.

The driver of the front-end loader nervously maneuvered his machine out of the field and set off for home. He wanted to be out of there as soon as possible.

As Larry and I walked back to the barn we saw Francis, Heather, and Jackie each carrying a lamb to the area in front of the media.

"Where are our rights?" Francis asked in a loud voice. Larry and I watched with baited breath.

"This is not justice. Where are our rights?" he repeated. "How can the government steal our property? How can they take perfectly healthy animals from a small family farm?"

He continued, "You are watching a theft take place. You are watching a crime being committed against an innocent family and innocent sheep!"

Everyone was silent. All eyes were on Francis and the girls.

I felt tremendous pride and pain. Here were our teenage children experiencing the most difficult event of their entire young lives, yet making sure their voices were heard and handling the situation with grace.

"I don't know why anyone in the United States would want to have a farm if this is what the government does to them. I have lost all my faith in the agricultural industry," Francis said loudly. "We have no rights!"

"Mrs. Faillace?" It was Bill Smith calling and motioning me into the barn.

"Yes?"

"We need to start loading the sheep. Could you please remove the llama?"

I called to Mary Jo, my tiny, redheaded, eight-year-old niece.

"Mary Jo, could you take Freddie to the rams' barn?"

"Sure, Aunt Linda."

She walked into the paddock, stood on tiptoe, and grabbed Freddie's halter when he leaned down to nuzzle her. Freddie loved this vivacious, beautiful little girl. She was always quick with a smile and sometimes had treats in her pockets. Holding his halter, she walked him out the solar barn, past the group of stunned grown men, down to the rams' barn. I followed her. Another group of USDA officials were standing by the rams.

I looked down at the loafers they were wearing.

"Where are you from?" I asked.

"Down south . . . North Carolina . . . A place that never gets this much snow!" were the responses.

"The USDA had to bring people from the South to do this job?"

They just looked at the ground.

"You should do some research. There is nothing wrong with our sheep. Please," I begged, "learn what the USDA is doing. These are the only sheep in the world ever accused of having mad cow disease. And they are perfectly healthy. You are helping to kill innocent sheep."

They quietly listened, but no one dared respond.

Larry and I stood numbly behind the tape, surrounded by family, as employees from the Vermont Department of Agriculture lured our beloved animals out of their warm, cozy home and onto the cold trailer that would take them to their death.

I covered my mouth as the tears ran down my face, staring in horror as one by one each of our animals ran past me.

At one point a protestor yelled "Wolf!" at a federal agent, implying the agents were wolves in sheep's clothing, stealing our animals.

However, the federal agents thought this call was sign of an attack and immediately became defensive. Agents grabbed at their weapons, not taking them out, but wanting them accessible. One agent rushed toward the protestor who shouted but fell into a snowdrift up to her thighs. The agent grabbed the nearest tree to try to pull herself out, but all the snow that had accumulated on the branches then fell on top of her. Two other agents came to her assistance and pulled her, cursing, out of the snow.

I insisted they load the ewes and lambs last. Brad was still holding Kanga, and I asked him if he wanted to carry her to the truck.

He stood silently for a moment, looking first at Kanga, then at Mary Jo.

"Would you like to carry her?" he asked.

Mary Jo nodded, too choked up to say anything. She took the beautiful, happy lamb into her arms and proudly carried her out of the barn.

The eyes of the world were now on Mary Jo.

One of the most famous photos of the entire battle was taken that day, as Mary Jo slowly carried Kanga to the trailer in tears, while Jackie and Hanna followed behind her with a plastic jug full of milk replacer—enough sustenance to last until the USDA killed her.

A few moments later Francis walked out of the barn, and in his arms was Moe.

I have never seen Francis cry like he did then.

I walked with the two of them to the trailer and sobbed as we put Moe in the pen with Upsala and her lambs. Moe immediately stuck his head out the small opening, baaing, wanting to know why I was not with him. He had never traveled before. He had never been away from me before.

All around us protestors were yelling, some were chanting, others just stood quietly.

"It's time to close the trailer up now," a USDA official told us.

Another USDA official walked around the truck with a bucket of bleach. He stopped by each tire and sprayed the outside of it. Once the trailer was securely closed, the driver drove the vehicle a few feet, stopped, and the tires were sprayed again. I have no idea what they thought they were accomplishing. I'm sure it was just for show. The truck was parked on a paved road and never even drove onto the farm.

Protestors blocked the path of the vehicle, so it was forced to move at a snail's pace. Larry and I walked beside the trailer, and I kept petting

Moe. As long as my hand was on his head he stayed quiet, but as soon as he didn't see it, he started to cry.

At the four corners, the protestors stepped aside, and the livestock trailer was allowed to proceed.

Moe started to bleat loudly, but his tiny voice faded as the truck drove off into the falling snow.

23

CONTEMPT OF COURT

Life was too busy to even think, let alone have time to feel. Heather organized our days into fifteen-minute time slots and scheduled interviews from 6:00 A.M. to midnight. Between the phone interviews, we met with reporters at our home. Friends and family filled the house with love and warm meals. Edie was busy trying to arrange to "send the family someplace warm." One day when I was on the phone with a live radio show out of Providence, Rhode Island, and Larry was talking with a reporter from Belgium, a woman walked into our apartment with a huge bouquet of flowers. Crying, she set the flowers on the counter, then sadly waved and left before either of us could thank her. Phone calls, e-mails, and letters flooded in from around the world as people sent money, gifts, and their sympathy. The outpouring of love was incredibly comforting.

Former USDA employees, elected officials, veterinarians, and scientists sent supportive letters; many were filled with research papers and helpful suggestions for proving the sheep were free of disease. A woman from Rutland, Vermont, sent us $100 and a note that read, "I want you to know I am thinking of you each day. Don't give up the fight and tell Secretary of Agriculture Glickman to drop dead." A woman from Utah sent $500 and a beautiful framed painting of Jesus surrounded by lambs.

Sally from Milton, Vermont, wrote, "I am writing this letter just to say that my thoughts and prayers are with you. You do not know me, but I feel that I know you, not only from watching your tragedy unfold on the television and in the newspapers but because for twenty some years I bred and raised sheep.

"I had to give up my farm and flock due to illness in our family. I truly know the loss you are feeling but cannot imagine the feelings of helplessness you must have gone through. I know there are no words that can give you peace or for that matter slow the grieving. I do pray that you and your little family will be able to dry your tears and lift yourselves above the pain and heartache. You are walking through a very dark valley, but the sun will shine and you will see the top of the mountain. God bless each of you."

A letter from Albany, New York, read, "I am a senior citizen in my sixties on a fixed income, but when I read the article in the *New York Post* my blood boiled. It's hard to believe our government has changed so where people's rights are stepped on. I wish I could say or do something to make you and your family feel better, but I am just one voice and so outnumbered by a government out of control. I do hope your son will be okay. There are just too few young people who have heart and really care. Keep your head up and don't ever give up. I just hope this note gets to you. May God bless you and your family, Carla."

The envelope was addressed to "The Faillace Family, Warren, Vermont" with no zip code or street address, as were dozens of others we received. Some were just addressed to "The sheep people in Vermont," and our post office still managed to deliver them.

Helen from Brooklyn, New York, wrote, "I cry as this is being written. There was a time I was proud to be an American. When Gestapo tactics are used to return a small child snatched from the ocean to a dictator and flocks of innocent sheep are stolen from hardworking farmers, disgust fills me. It is my belief that your sheep were taken for other purposes and not all will be slaughtered. The very worst is that the truth about your sheep's health will never be known. Please know that your family and precious animals are in my most ardent prayers. May whatever hardship this is causing financially be over soon. Thank you most of all for being wonderful animal people."

A sheep and goat farmer from New Hampshire wrote, "Perhaps it will comfort you to know that some of us have been following your struggles to save your sheep, wishing we could help, but unable to do much more than feel your pain. As senseless as all this seems to be at this point in time, one day a reason will reveal itself."

This sentiment was often echoed. "You aren't alone. There are many of us out here on your side. We share your sorrow but hope we can rejoice when you emerge the victor!" from Freda in West Danville, Vermont.

We sent more letters to Ann Veneman and the USDA lawyers, pleading with them to spare our sheep, but they never responded.

Many people called and shared horror stories of losing their animals. A chicken farmer from New York had his flock accused of having a disease, yet all the independent tests came back negative. But the USDA "accidentally lost" the test results and seized the flock; another farmer in Vermont had his cattle seized; a sheep farmer from the Midwest had all her sheep seized and killed when a ram she borrowed to breed her flock tested positive for scrapie. The USDA destroyed all of her sheep (which were healthy), yet did not destroy the flock where the ram was from.

A couple from Washington called and begged Larry and me to preserve our relationship. Their friends and neighbors had imported deer from China with the permission of the USDA. A few years later, after their deer business was well established, the USDA claimed the animals had to be destroyed for fear they could be "susceptible" to chronic wasting disease. There were no symptoms, no positive tests, just the fact that the animals were imported. (And, oh yeah, China didn't have chronic wasting disease, only the United States did.) The couple attempted to fight the USDA, but the fight was too hard, lasted too long, and the USDA destroyed their animals, destroyed their business, and the stress from it all eventually destroyed their marriage. "Promise us you will take care of each other and your family," they pleaded to us over the phone.

All these stories made me even more determined to fight back. It was time for small farmers to stand up against a bureaucracy out of control. We had to do something. As Larry told the *Boston Globe* the day after the seizure, "This is the beginning of a new chapter of our life and of getting the truth out to the public. The USDA can destroy our sheep, but they can't destroy the truth."

On the Monday after the seizure, Dr. Bill Smith called. He was the only USDA employee to apologize sincerely for what happened. "I'm sorry," he had said at the end of a meeting we had at the Vermont Department of Agriculture. "I never thought it would come to this."

On this particular day he needed information about the sheep semen. "How many straws of semen do you have from the sheep?" he asked.

"Somewhere between two hundred and four hundred," I said. "I don't know the exact number, but I have it on a receipt and could get it for you."

"That's not necessary," Bill said. "Do you have any embryos?"

"No, I wish we did," I said sadly. "What does the USDA want with the semen?" I asked. "There's no evidence of any TSE being transmitted through semen."

"I know," said Bill. "I think they just want to finalize things. To know exactly what exists."

If the USDA took the semen we would have absolutely no genetics left from our animals—no East Friesian, Charollais, or Beltex. There would be no possibility of starting over. At least with the semen we could slowly start again by inseminating American ewes.

Bill asked about the children and spoke of his own. He remembered seeing the rams with Francis and was impressed with his connection and dedication to them. "He knew all their names and their idiosyncrasies."

We talked about the children soon going to college and how life was passing too quickly. Before we hung up I asked him again, "You don't think they will try to take the semen, do you?"

"No," Bill said. "There should be no problem with the semen."

No sooner did I hang up, when the phone rang again. "Is this the Faillaces?" the woman on the other end asked.

"Yes it is."

"You don't know me, but I have a second home down the road from your store. My name is Wesley Porter, and I have been following your story. I was in California when they took Mr. Freeman's sheep and in Arizona when the USDA grabbed yours. Well, my heart goes out to you, but especially to your children. I watched them raise and care for the animals. I used to see Heather milking the sheep when I would drive by, and they were out in the fields every day. It was crushing to watch them have the animals taken. I saw the entire thing on CNN while I was in the airport.

"Now I know it's not much," she continued, "but I have a home in the Dominican Republic, and it is available from now until April 15 and your family is welcome to use it."

I caught my breath. "Are you sure?"

"Oh, it's nothing. Nothing fancy. It's just a villa which should be big enough for all of you."

"That's extremely generous of you, Wesley," I said.

"It's the least I can do. With all the children have gone through, nothing against you and Larry, but my heart breaks for your kids, and

I would be very happy if I could do something. If you go to the Dominican Republic, you can get away from all the interviews and be somewhere you're not recognized. Plus it's a beautiful country."

I thanked Wesley profusely.

The next day Larry and I gave testimony at the statehouse in Albany, New York. Assemblyman Felix Ortiz and Assemblywoman Margaret Markey of the New York Legislature invited us to address the task force on Food, Farm, and Nutrition Policy about Emerging Food Related Diseases, and specifically to talk about BSE and whether the actions the USDA was taking were strong enough to protect the American consumer. Ortiz and Markey sponsored the public hearing to discuss TSEs and the New York food supply. According to Felix Ortiz, "As the new chair of the task force on Food, Farm, and Nutrition Policy, I want all New Yorkers to have a healthy, affordable, food supply. I don't want to discover cases of this deadly disease [BSE]; I want to prevent it. At the same time I would like to help increase the sales of New York–grown and –produced foods. One of my goals from these hearings would be to develop a program to certify that New York animals are the safest in the world so that New York farmers can successfully market those meat products to consumers here in New York State and around the country. I have requested $100,000 in the state budget for a pilot project to test cows slaughtered in New York for BSE."

Larry and I spoke for ten minutes each. We spoke of the fact that the government needed to work with farmers, not destroy them, if there was any hope of eradicating TSEs. We were an example of a farm that had followed all the rules, particularly as they pertained to scrapie:

- We could verify all sources of feed for our sheep and their ancestors
- We started the scrapie program in Vermont and enrolled in it
- We promoted the scrapie program to other sheep farmers
- Our animals had been monitored for eight years, and there were never any clinical symptoms
- We used organic/sustainable practices
- We were willing to work with the USDA to set up additional testing programs

• • •

The government testing programs were aggressive in sheep for scrapie and theoretical BSE, passive in cattle for BSE and humans for CJD, and practically nonexistent in deer and elk for CWD. New York had a unique opportunity to assure consumer confidence by working with farmers on an "approved" testing program. If the farmers understood exactly what tests would be used and what would constitute a positive or negative, and what the consequences for their farm would be, they would be much more cooperative. And consumers would buy New York beef over others, confident in its safety.

As soon as we finished, a local television crew asked to interview us outside the statehouse. Once the interview was over, they invited us into their mobile van to watch the editor piece together the story. It was the first time I saw footage of our seizure, and I sat in shock, watching our animals chase the grain bucket onto the trailer. Instinctively I covered my mouth. I could feel the tears start. Larry leaned over to hold my hand, and I clung to his tightly.

Back in the building we called the children. Everything was fine. The phone was still ringing off the hook, but the calls could wait until tomorrow, except for Davis. Would we please call him?

Davis was laughing when he heard me on the line. "The USDA tried to hold you in contempt of court today," he said.

"What?!" I cried. "What did I do?"

"Nothing. That's why it's so funny they targeted you of all people." Davis was having a hard time getting the words out, he was laughing so hard.

I didn't find it funny. What did it really mean to be in "contempt of court" anyway?

"They claimed you were aggressive with Bill Smith and refused to tell him where the semen was," Davis said.

"What?! . . . " I said, sputtering.

"Did you give him hell?" Davis asked conspiratorially.

"Of course not!" I practically yelled into the phone. Larry reminded me people were still giving testimony and the halls were echoing.

"I know you didn't," Davis said, still chuckling. "I think I know you better." He explained how the USDA lawyer faxed a request at 11:14 A.M. to disclose the exact location of the semen. The USDA gave Davis until 3:00 to answer the request. Davis was unable to respond as we were in Albany and unreachable. So the USDA claimed we were not cooperating and withholding information.

As a result of this, I had to file an affidavit recounting the phone call with Bill and reminding the USDA that when they asked about the semen previously we had responded in writing on March 14. The judge threw out the USDA's motion to hold me in contempt of court.

On April 6 the entire family, and Francis's close friend Kortney, went to the Dominican Republic on a trip completely funded by the community and money people from around the world had sent.

Wesley's "nothing fancy" turned out to be an amazingly beautiful home with three bedrooms and three baths. Nadie, Wesley's maid, did all the cooking, cleaning, and laundry. We had never been to the Caribbean before, and the warm sea air, fragrant flowers, and sunshine soothed our souls.

The first three people we met had either heard our story from Wesley or had seen us on CNN. One of these was the caretaker for the property, and he owned a dairy cow farm high in the hills, about an hour and a half's drive from our villa at Caso de Campo. "I'd like to take you to see my farm this weekend," he suggested. "And I would like for Heather to milk a cow. I think it would be good for her."

Larry agreed, with a grin on his face. There was no way he would tell Heather ahead of time what the plan was, or she would spend the entire time fretting about it.

Saturday morning the caretaker and his wife and two children arrived shortly after breakfast, and all six of us piled into their SUV for the bumpy ride into the beautiful mountains. When we arrived at the farm, two farm workers waved as they mounted horses bareback, rode out to the fields, and herded the cows into the barn.

As soon as we entered the barn, one of the workers motioned to Heather to join him milking. Heather looked at him in confusion and then glanced behind her thinking he must be pointing at someone else. The worker smiled a beautiful bright smile and pointed again at her. Heather shook her head. He continued beaming and shook his head to mean he wouldn't take no for an answer. She looked helplessly at Larry and me.

"Go ahead," the caretaker said. "It will be good for you."

Heather looked down at her shorts and halter top and back at Larry. He grinned. Carefully she made her way over to the cow, trying to keep her flip-flops clean. With a look of satisfaction at winning the silent battle, the worker bent down and demonstrated how to milk by hand into the bucket. Heather knew exactly what to do.

The rest of the morning and into the afternoon was spent leisurely watching the children ride horses around the incredible farmland and learning about the Dominican way of life through their food and traditions. The trip could have ended that evening, and we all would have been content.

The phone in the villa rang, and Francis answered, "¡Hola!" It was Marc Harmon on the other end. Marc and Jacki were staying at our house and taking care of Pélé and Freddie (the llama) and answering the phones for us while we were away.

"I'm trying to reach the Faillaces," he said.

"¡Hola!" Francis repeated.

"Could you please connect me with the Faillaces?" Marc asked a little louder.

Francis saw his opportunity. Francis has an amazing talent for imitating people and acted like he was speaking Spanish and put Marc on hold. Marc waited patiently.

A few seconds later Francis asked, "¡Hola! ¿Puedo le ayudo?"

Marc started to rattle off something in Spanish, but Francis was already at his limit.

"¿Qué?" said Francis.

Marc repeated himself, but this time more slowly.

"Ah!" Francis said and proceeded to speak very quickly with Spanish-sounding words.

Marc was flustered. He thought he was fluent in Spanish but decided to abandon it for English. "Please may I speak to Dr. Larry Faillace?" he said carefully and loudly.

"Saillace?" Francis asked.

"Faillace," Mark said, emphasizing the "f."

By this point the rest of us were laughing hard. It was wonderful to see Francis get his sense of humor back. Francis quickly turned and put his finger to his lips, signaling us to be quiet.

"¿Me casa?" Francis asked.

"¿Me casa?" Marc was confused.

"¿Faillace casa?" Francis asked again.

"¡Sí! ¡Sí!" Marc excitedly replied. Now he was finally getting somewhere. "¡Sí! Faillace casa."

"¡Sí! Faillace casa, uno momento," Francis said, handing me the phone before he burst into laughter.

"Hi, Marc!" I said, wiping the tears of laughter from my eyes.

"I'm so sorry, Linda," he began. "I tried to make this call short. I didn't want to run up your phone bill, and I got caught up in the phone system. The first guy put me on hold and then the second guy . . . well; I couldn't keep up with him. He spoke Spanish faster than I've ever heard."

At that point I couldn't stop laughing. "It was Francis," I said.

But Marc was still too busy apologizing to realize what I just said. "I kept asking for you, but their accent is really thick."

"Marc," I said, "it was Francis. That was Francis on the phone."

Marc was silent for a moment. Then he said, "I didn't know he spoke Spanish."

"He doesn't!" and with that the laughter began again.

"Is everything okay?" I asked when I was able to compose myself.

"Everything here is fine," Mark said. "Pélé and Freddie are good. But you received a phone call from a Roger Ives. Something about semen? He seemed upset, enough so that he wants you to call him."

Marc gave me his number. I thanked him and apologized for Francis's joke.

"It's okay. I'll get him back," he said.

"Join the list," I said, laughing. "Join the list."

Roger Ives was shook up. "I'm sorry, Larry," he kept saying.

"It's okay. It's okay," Larry said, unsure what Roger was so upset about. "What's up?"

"They showed up at my house," Roger said.

"Who?"

"The thugs for the USDA."

"What do you mean?"

"My wife and I were getting into bed around 9:00 when the phone rang." Roger was in his early eighties and extremely fit. For the past fifty years he had traveled the country collecting, freezing, and storing livestock semen. His company, Soligenics, was the best on the East Coast.

"I answered it, and it was some guy asking about the semen from your rams. He said he wanted to pick it up," Roger continued. "I told him I would have to talk to you first, and then we could make an appointment.

"He got aggressive with me and told me he wanted to pick it up now. I told him not to be ridiculous, it was late at night and I needed to talk to you first. But he insisted, and when I told him we were going to bed, he informed me he was at the end of my driveway! They were watching us!"

"Bastards!" Larry swore quietly. There was no reason Roger and his wife should have been harassed like that.

"But I refused to be pushed around. I told them they could come back the next morning," Roger continued. "Next thing you know my wife sees our house on the CBS news. Do you believe it! That really got us! How dare they!"

"I tried to get ahold of you, but they were there first thing and took everything. I'm sorry."

"Do you know who it was?" Larry asked.

"A guy in a green SUV," Roger said. "I think his name was Finn."

Larry thanked him for everything he had done, and before Larry hung up, Roger said, "Sure is a waste to see those genetics destroyed."

24

EMPTY BARNS

Reality hit hard, harder than I ever expected. When we returned home, the barns were empty and the fields were empty. Gradually the phone stopped, and the e-mails and letters slowed down to a trickle. The only thing full was our home. Our friends, Marc, Jacki, and their daughter, Laci, were looking for a place to settle, and we offered for them to stay with us in the interim. Then my aunt joined the crowd and moved in for a few months. Despite being surrounded by friends and family, nothing anyone could say could take away the pain.

When we lost the sheep, we not only lost our source of income and way of making a living, we lost our friends and pets, and the children lost their jobs, a huge part of their identity. Everyone in the family had put their heart and soul into making our business successful and now it was gone.

We all responded in different ways. Francis became angry and withdrawn, snapping at anyone who attempted to talk about the sheep, agriculture, or our future. Heather rebelled. It was hard to know what we could attribute to her being a fifteen-year-old and what was the result of losing her sheep. Jackie withdrew, refusing to talk to anyone, refusing to walk anywhere near the barn. Larry stayed strong. He needed to hold his family together. His family, which once sat around the table every evening excitedly making plans for the future, could now barely enjoy a meal together.

And I sunk into a depression. A deep depression. My nights were filled with visions of Moe calling to me, of being trapped in quicksand unable to get to him as men in black hauled him off to an incinerator. I dreamed of walking Roxbury Mountain Road and seeing our sheep lining both sides of the road, in the gullies, dead. I searched each animal, hoping one of them was still breathing.

During my waking hours I struggled with the past. How could I have prevented my children from feeling such tremendous pain? Were we wrong to give them so much responsibility? What did our future hold? It was with that question that the blackness took over. Nothing. The future held nothing. What could I offer my children? What could I offer my husband? My days were spent crying. I lashed out at the family over the smallest thing and then retreated into the bedroom in shame and sorrow.

Davis reported to us that our case was declared moot in second circuit court because the USDA told the judge there would be no further interaction with us. The case was about the sheep, and the sheep were dead. The USDA had nothing else to do to us. They took the sheep, they took the semen, they stole our dreams. There was nothing left.

One afternoon, Ann Veneman was on CSPAN talking about the Food Safety Summit. Veneman reassured Americans that the "new [food safety] regs have to be based on science that is both thorough and sound. Speculative science is good for grabbing headlines, but we need good solid science to base our regulations and policies on."

"Turn it off!" I told Larry angrily. The USDA learned that stealing our sheep and killing them for a disease that didn't exist was "speculative science . . . good for grabbing headlines."

"Let's hear what the other side has to say," Larry told me as Alison Beer, editor for *Food Chemical News* spoke. "Politics are ruling more of government agencies, and decisions are made less on science and more on politics," she said.

Before our battle with the USDA, I was unaware of Washington's "revolving door" policy. Lawyers, advocates, and lobbyists for large corporations or special interest groups leave their positions to work for the U.S. government, where they are able to push through legislation for issues they worked on and then return to their positions, perhaps to do another round at a later date. The doors have been revolving for many decades, but during the George W. Bush administration they practically came off their hinges, with more than one hundred top officials swinging through the doors.

Michael Taylor and Linda Fisher were prime examples. Michael Taylor was a lawyer working for Monsanto. Then he was hired as a legal advisor for the FDA's Bureau of Medical Devices and Bureau of Foods, then back to Monsanto, and finally as executive assistant to the commissioner of the FDA.

Linda Fisher was assistant administrator of the EPA's Office of

Pollution, Prevention, Pesticides, and Toxic Substances for ten years. She knew the laws inside and out, so she was hired by Monsanto as the deputy administrator for their Washington lobbying office. She went on to serve on the USDA's advisory committee on biotech foods. Then it was back to Monsanto, where she was vice president of government and public affairs. And in 2001 Bush nominated her and she was approved for the second-ranking position at the EPA.

As noted in Philip Mattera's report "How Agribusiness Has Hijacked Regulatory Policy at the U.S. Department of Agriculture":

> In its early days, the United States Department of Agriculture (USDA) was dubbed the "People's Department" by President Lincoln, in recognition of its role in helping the large portion of the population that worked the land. Some 140 years later, USDA has been transformed into something very different. Today it is, in effect, the "Agribusiness Industry's Department," since its policies on issues such as food safety and fair market competition have been shaped to serve the interests of the giant corporations that now dominate food production, processing, and distribution. We call it USDA Inc.
>
> The reorientation of USDA has been occurring over many years, but it has now reached a dramatic stage. Thanks to its growing political influence within the Bush Administration, Big Agribusiness has been able to pack USDA with appointees who have a background of working, lobbying, or performing research for large food processing companies and trade associations. Conversely, there are virtually no high-level appointees at USDA with ties to family farm, labor, consumer, or environmental advocacy groups.

Eric Schlosser, author of *Fast Food Nation*, agrees: "Right now you would have a hard time finding a federal agency more completely dominated by the industry it was created to regulate."

Carol Tucker Foreman was the USDA assistant secretary under President Carter, and she said, "Whether it's intentional or not, USDA gives the impression of being a wholly owned subsidiary of America's cattlemen. Their interests rather than the public interests predominate USDA policy."

Examples? Meat and livestock organizations gave $623,000 in campaign donations in 2000. According to the Center for Responsive Politics, Al Gore received $23,000 and George W. Bush received $600,000. Dale Moore, former executive director for legislative affairs for the National Cattlemen's Beef Association, became chief of staff for Secretary Veneman. Elizabeth Johnson became one of her senior advisors. Johnson's former position? Associate director of food policy at the National Cattlemen's Beef Association.

Secretary Ann Veneman's previous position? Former director of Calgene (part of Monsanto, which is now part of Pharmacia). Calgene promoted the Flavr Savr tomato—the first genetically engineered food. Charles Lambert spent fifteen years working for the National Cattleman's Beef Association and in 2002 became the USDA undersecretary for marketing and regulatory programs. Assistant Secretary for Congressional Relations Mary Waters was formerly a senior director and legislative counsel for ConAgra Foods, and the list goes on.

Ann Veneman continued talking on CSPAN about how great the USDA was and how the United States was free of BSE because of their "stringent firewalls in place."

I walked out of the room, disgusted. I didn't want to hear any more.

The cover story for the April 30, 2001, *Wall Street Journal* featured Dr. Mary Jo Schmerr and her blood test. Dr. Schmerr claimed "her superiors were obstructing her research by limiting her travel and collaboration with other scientists, and even belittling her." Dr. Paul Brown of the National Institutes of Health (NIH) criticized her blood test and said, "It's the worst test that I have ever seen in forty years of working in the laboratory from the point of view of reproducibility and standardization. If you can't reproduce an experiment, the experiment cannot be considered science."

Dr. Robert Rohwer also criticized her test when a joint blind testing program demonstrated some of her findings weren't consistent with his. But he also acknowledged that he was a competitor as well as a collaborator as he was likely to be "a significant player" in a company being formed to develop a different test. (Rohwer had a five-year, $10 million NIH contract to develop a diagnostic test for TSEs.)

Another researcher with the American Red Cross publicly stated that Dr. Schmerr's test could not tell the difference between healthy and diseased blood. Dr. Schmerr admitted, "This is a test in development. It needs to be improved. It will not be 100 percent accurate."

The *Wall Street Journal* article also pointed out, "Dr. Murray [of the USDA] recently required Dr. Schmerr to do a blind test of the procedure on about 200 sheep-blood samples. Dr. Schmerr says this wasn't appropriate because the technique is a work in progress. The blind test produced numerous 'false positives.'"

The USDA knew exactly where to go when it wanted to produce false positive results.

On May 8, 2001, the Vermont Department of Agriculture sent a letter informing us "the remainder of the quarantine on the subject sheep is hereby lifted."

I sat on our back deck and read the notice. Part of me wanted to tear it in half or shred it into a thousand tiny pieces. I looked out over our pastures, the grass getting taller with each passing day. The fields should have been filled with sheep; Heather starting to milk soon, and Jackie preparing to make cheese. The wind played with a piece of the metal roofing from the three-sided sheep shed that was now in pieces, dismantled on the ground.

More than a month had passed since the seizure, and still there was no word on test results or on compensation. I watched a squirrel run up a tree that bordered our lawn and the fields. It was the only living creature I saw. Whenever the USDA was asked by the media for the results there was always some sort of excuse, the best from Detwiler: "We have decided to up the biosafety level and build a special building for processing the sheep samples, in case it is BSE. We want to protect our lab techs." Protect the lab techs at National Veterinary Services Laboratory in Ames, Iowa, from BSE? That was the laboratory that performed all of the USDA's diagnostic testing for BSE since 1990. Did that mean Ames was not a safe place to test for BSE?

The wind picked up, and I pulled my jacket closer. There was still a nip in the air, winter letting out her last breath. I sat on the steps and stared across the farm where lambs once "spronked" in the fields, and it made me think of Moe. He was born in that field. Tears streamed down my face as I recalled the chaos and excitement of that lambing season.

. . .

The USDA sent us a three-page letter in June for "Cleaning/ Disinfection and Premises Remediation Recommendations for the Faillace Premises."

"What the hell is this?" Larry asked Davis.

"More bullshit from the USDA," Davis agreed. "They told us in court they were done. Nothing else to do with us. That's why the judge declared it moot. And now this!"

"Can they legally do this?" Larry asked.

"Legally? I don't know. But you can be damn sure if we fight them, something unusual will 'appear' on the test results," Davis said.

All wood from the barn was required to either be sealed or incinerated. We were to clean, wash, and disinfect (with a 2 percent chlorine solution) everything else that was physically possible. "If they think a 2 percent chlorine solution would denature any possible abnormal prions," I said sarcastically to Larry, "we could have just run the sheep through our hot tub every night!"

Our solar barn had to be disassembled and that meant moving Freddie. I was not the only one who was depressed. Freddie refused to come out of the barn, even to graze. He spent day and night laying in the barn, waiting for his flock to return. No amount of prompting or promises of "treats" could convince him to leave the barn. So we fenced in a paddock behind our house to put him closer to the family. It helped a little. Then we brought two American sheep to keep him company. Freddie began eating more. Larry had always wanted to raise geese, so we ordered goslings through the mail, raised them in a tub in the living room, and soon Freddie had a dozen noisy companions. Freddie loved them. Now he would have new friends to guard.

We were in the process of cleaning the farm when the USDA sent us a letter regarding compensation. Again they said they would pay "fair market value" for the sheep and germplasm seized and destroyed and said "in an effort to fully determine fair market value for the sheep and the semen, we request that you submit any purchase or sales receipts, milk production records, purebred registrations, progeny records, cheese production and sales records, records on the number of ewes by lactation cycles, and other documentation that would assist in our determination. In addition, we invite you to present in writing any positions you have with regard to appropriate compensation, with documentary or other factual support for those positions to Dr. Bill Smith by July 1, 2001. After you have submitted this information and we have had an opportunity to review it, we would be willing to schedule a meeting with you to discuss the matter in more detail." It was almost identical to the list they had sent us in November 2000.

Putting together all the information once more for the USDA did nothing to improve my frame of mind. Larry and I spent day after day gathering production records, organizing all the pedigrees, and creating a three-inch, three-ring binder full of documentation about our farm.

I read online that the sheep had been killed. Papers from Iowa and Missouri interviewed Detwiler, and she explained in great detail how the animals were seized because they "were either infected with scrapie or mad cow disease," and if they were infected with BSE "the Vermont sheep would be the first domestic livestock other than cattle to come down with mad cow disease outside of a laboratory." She continued with gruesome details of their destruction: "The animals were euthanized and tested and all the sheep were dead and their carcasses were consigned to 'the digester'—a hot lye bath built specifically to deal with animals infected with prion diseases."

The USDA used our situation as an opportunity to pressure the remaining imported cattle owners to give up their animals. In early April, a farm in Texas surrendered sixteen imported German cattle that were tested by histopathology at Ames. All were negative, and the test results were announced less than three weeks after the animals were killed. From a *Times Argus* article we learned that "one Texas producer imported four head of the German animals. He sold one animal for diagnostic purposes and agreed to sell another for testing in about forty-five days, but does not want to sell the remaining two animals. The remaining animals are under restriction and surveillance by a USDA field veterinarian. The owner is instructed to report any health problems so the animals can be re-examined immediately."

When the USDA attempted to gather all the cattle originating from Britain, two Vermont farms refused to sell their animals. According to an article by John Dillon, "the government [USDA] encountered a similar streak of Vermont stubbornness among the local cattle farmers. One operation steadfastly refused to relinquish its six prize belted Galloway cattle despite government pressure. Another farm was reluctant to sell its two imported Scottish Highlanders but finally acquiesced to the USDA. 'There were some [farms] that didn't sell right away, but after a while, everybody else sold,' said Dr. Linda Detwiler."

The article also said, "Government vets inspect the cattle every three months for any sign of the fatal brain illness . . . and semen from the cattle can be used for artificial insemination."

The farm owner "resisted selling the animals to the government because their value as breeding stock far exceeded the price the government was offering."

"Good for them," I thought. At least they were able to hold on. There was absolutely no reason why their animals should be seized either. What made me angry was the fact that the imported cattle were immediately tested and reported negative. It was now more than three months since our animals were stolen, and still no test results.

Even President George W. Bush condemned our animals. From an article in the American Sheep Industry Association newsletter: "In remarks to farmers in Montana this week, President Bush said he was pleased with the actions to remove the flocks from Vermont."

Praises were raining down for Detwiler. The *Washington Post* published an article about Detwiler keeping mad cow disease out of the United States. "She was at the center of a bitter dispute over the March confiscation of two flocks of European-born sheep being raised in Vermont. Detwiler was convinced the sheep had to be destroyed because there was the possibility, however slight, that they were carrying the mad cow agent. But the flock owners and their supporters believed the government was overreacting. During the final court-ordered seizure, protestors carried signs denouncing 'Dr. Deathwiler.'

"For the no-nonsense Detwiler, raised on a New Jersey hog farm and enamored of farm animals all her life, the criticism hurt and led to some restless nights. Referring to the Vermont sheep, she said it is always difficult to 'depopulate' a herd. But living on a farm as a child, she quickly learned that sometimes individual animals had to be sacrificed for the general health of the group. 'I grew up with hog cholera, so I know how it feels to lose some animals to protect the others,' she said.

"In dealing with a danger such as mad cow disease, perception of risk can be as important as the risk itself. So part of the federal effort—and Detwiler's job—has been to explain time and again that mad cow disease is not present in the United States, and that if it was ever discovered, it would be quickly contained." (I had naively thought U.S. law allowed the USDA to deal with reality, not perceptions.)

Detwiler and the USDA were getting away with misleading the public about the woeful inadequacies of a system that was designed to protect the beef supply. The USDA was so proud of Detwiler they honored her with an award for "Heroism and Emergency Response" for dealing with our situation.

I sank even deeper into depression.

25

WHERE WAS LEON?

Honey, it's time for you to snap out of it," Larry said when he found me alone at home crying again in the middle of the afternoon. I had stumbled across a photo of me holding Moe. I turned and looked at Larry in anger. What I was feeling was too deep. It wasn't something I could turn on and off, something I could just "snap out of."

"You need to get involved with something, something to take your mind off everything," Larry said, attempting to move closer, but the barbs flying from my eyes told him to keep his distance. I felt like a caged animal, trapped in my emotions.

"Why don't you reopen the store?" The store had been open sporadically between October of 2000 and January of 2001, but by the end of January we were so overwhelmed with dealing with the USDA that running it was no longer an option. We shut the doors. Now it was June, and the store remained closed.

"What about getting some sheep once we get the money from the USDA? We could get some Texels from Ron in Connecticut," Larry suggested.

That was the wrong thing to say. I thought of Moe and started sobbing.

"C'mon, Lin," Larry said comfortingly. "We've got to do something. I can't stand seeing you like this. Maybe you should write a book. Maybe if you put your feelings down on the page it will help you heal."

I looked at him like he had two heads. The last thing I wanted was to relive those memories. I shook my head no.

"All right then," Larry said, taking charge when he saw I was being immoveable, "let's go for a walk. You need some fresh air and to get outside."

I was too worn out to argue and followed him. Instead of heading toward the barn, like he usually did, Larry turned toward the store, and I tagged along, as did my black cloud. He slowed so we could walk together and took my hand. "Put your head up," he said gently. "You spend too much time looking at the ground these days. Look at the mountains."

I looked up at the gorgeous view and the heat of the sun on my face felt nice. Larry turned into the schoolhouse parking lot. I pulled my hand away. "Where are you going?" I demanded.

"We are going for a walk," he said, grabbing my hand back and pulling me with him. He walked me into the schoolhouse. When he opened the door to the market he acted like he was searching for something and left me standing in the doorway. Tears rolled down my face, and I surveyed the shelves now empty with the exception of a few nonperishables, everything covered with a thick layer of dust.

"Why did you bring me in here?" I asked with hurt in my voice.

"Look around you, what do you see?" Larry pressed gently.

"I see one more thing the USDA destroyed," I said. "Another dream gone."

Larry laughed, "You know what I see? I see potential. They didn't take this away from us. We still have it. The store is still here, the family is still here."

I wondered how he could have any optimism left. Mine was gone. But before I could question him about it, he said, "I think it's time for another 'holistic resource management' meeting with the children."

I walked over to one of the shelves and ran my finger through the dust. Maybe Larry was right. Maybe it was time to start over.

In 1999 a friend of ours taught a course on Holistic Resource Management (HRM) at Rootswork. HRM is a model that considers people, economics, and environmental health as inseparable. It is also a tool for planning the future. Larry and I had adapted our own version and found it to be helpful in making decisions.

That night Larry announced over dinner that the next afternoon the family would have an HRM session. Groans were heard around the table. These "sessions" tended to last three or four hours and were not high on the children's list of favorite ways to spend an afternoon. But regardless, at 1:00 the following day we were gathered with our pens and paper at the dining room table.

"Okay," Larry began, "I want to go around the table and for you each to give your opinion about getting more sheep—"

"Wait, wait," I interrupted. "I think we should first write down what we each would like to do once we get paid by the USDA." I wanted to know how the children felt, and I thought they would express themselves more truthfully on paper.

Larry smiled. My interruption was a sign to him. A good sign. A sign of hope. "All right. We will do what your mother said," he told the children.

At first Larry was the only one writing. The children and I sat there, staring off into space, thinking, not making eye contact with each other. Then Heather bent down to start writing. A few seconds later so did Francis and Jackie. The only one not writing was me.

I sat there, still thinking. What did I want to do? When we got paid by the USDA we could start over doing anything our hearts desired. What did mine desire? I could hear the clock ticking on the wall and watched as everyone else was now focused on their writing.

Did I want to have more sheep? Should we reopen the store? What about this idea Larry had about writing a book? My parents were writers and I loved writing as a child, but with motherhood and marriage the time and passion for writing had faded.

I looked over at Heather's paper. What did the children want? She pulled her paper closer and covered it with her arm, giving me the look someone would give a cheater attempting to see the answers. I couldn't help but stare as her pen flew across the page. It reminded me of taking our first trip from England back to the United States. Heather was in first grade and Jackie was in kindergarten. They were each given a journal to write about their trip. Heather wrote voraciously from the moment we set foot on the plane. She wrote so much, we had to buy her a second journal. There were so many thoughts coming out of that tiny body and young mind.

Jackie ended the trip with a single paragraph. "We got up. I got dressed and ate breakfast. We got on the plane." The difference was that Heather would write and write without any regard to style or neatness. Jackie's paper was threadbare from constant erasing. Every single letter was perfect. Her writing looked like something from a book teaching children how to shape their letters.

"Hon, you need to write, too," Larry said, startling me. I looked over at him and nodded, "Okay. Okay."

I took the cap off my pen. What *did* I want? I wanted my family to be happy. But how? Was it with sheep? I began writing. I knew that I wanted to work together as a family. We did it before, and that was the best part of the farm. I wanted to do something that was related to agriculture, maybe gardening. I wanted to do something for our community—to repay everyone for their kindness. When my pen touched the paper, the thoughts flowed, and I began writing.

"Mom, can we get started now?" Jackie asked a few minutes later. I looked up and noticed everyone else was finished.

"Sure, Honey," I said, putting my pen down on my paper, now covered with ideas.

We went around the room and everyone read their thoughts. The children were unanimous: They did not want to have sheep yet, it was too painful. What they did want was to reopen the store. Heather had visions of the store being open full time, having a booth to sell cheese at the local farmers' market, and turning the garden shed into a café. Jackie also wanted to see the store open but was mostly interested in making cheese again. Francis wanted to see some of the construction jobs completed including the porch on the schoolhouse and the cheese aging cellar, and he threw in a few suggestions about "getting even." We laughed and told him this time was for creating, not destroying. Larry wanted to do it all: get more sheep, open the store, have wonderful organic gardens, and make cheese. I just wanted my family to be happy.

We made a list of all our resources, including physical (for example, truck, barn, cheese facility, store) and more esoteric (for instance, liking to interact with people, teaching abilities, community support). Four hours later we had the start of a plan.

We cleaned the store, spent hours and hours mortaring the aging cellar walls, and Jackie and Larry bought more milk from a local cow dairy farm and made cheese. Jacki Harmon spent days refinishing the floors and shelving in the store and sewing curtains. My aunt loved to weed and with her own pair of gloves and tools spent many happy hours in the gardens, bringing them back to life. The USDA was coming to inspect the farm in August and we had to dismantle the barn, but now it was not as painful, because we were busy building our future.

On August 28, 2001, Wayne Zeilenga and Bill Smith arrived at the farm with Dr. David Taylor of Scotland and Dr. Richard Race from the NIH Rocky Mountain Laboratory in Montana to inspect our farm and offer recommendations regarding "cleanup." By this point, the store was now open, cheesemaking was in full swing, and I had some of my feistiness back. We invited Davis to join us, and he laughed as I verbally launched at Race and Taylor.

"So do you have the test results yet? It only takes between six hours and two days to perform all the testing. It's been five months and no test results. How are you going to make recommendations for a farm that all the animals tested negative? Isn't this a waste of your time? Why would the USDA pay for experts to fly over from Scotland and Montana to walk our fields in Vermont? Especially without test results?"

The media had hounded the USDA for the test results, and USDA spokesperson Ed Curlett kept saying they would be ready around the end of June. Larry filed a FOIA in early June requesting the test results. June came and went—no test results, no response to the FOIA. On August 3, Curlett announced there was a delay and "there's no predicted time frame for when they will be finished." He told the Associated Press, "NVSL has to build a high-security lab, resulting in the tests taking longer than expected." Yet when interviewed by Cory Hatch of the *Valley Reporter* the same week, "Curlett dispelled the reports that the USDA is building a new facility; he did say that the facility was purchasing several new pieces of equipment to aid in the testing process." Curlett said the USDA was taking their time "because we want to do the test correctly." So which story was right? And why hire people to inspect our farm without having test results?

Race and Taylor looked at me with a slight amount of fear in their eyes; they must have wondered about my sanity. But I refused to stop. "How come we had to spray a 2 percent chlorine solution all over?" I asked Taylor.

"Well, my research has shown that strong sodium hypochlorite solutions inactivate the infectious protein," he replied.

"If that's the case, how come the sheep had to be put through 'the digester'?" I demanded. He didn't have an answer.

Wayne and Bill had donned coveralls and were washing their rubber boots in a chlorine solution. Taylor and Race began walking toward the barn, dressed in their everyday shoes. I ran in front of them. "Where do you think you're going?" I demanded, blocking their way.

Wayne and Bill looked over and sighed. This was going to be worse than they thought.

"I will not allow you on our farm in those shoes," I said to Taylor, pointing at his dress shoes.

"And why not?" he asked.

"You're supposed to be the expert," I said with exasperation. "You are from Scotland and could contaminate our farm with foot-and-mouth. Change your shoes."

This was a problem. Wayne did not have a pair of rubber boots for Taylor. I asked Taylor for his shoe size, and Francis ran to the garage and got a pair. We stood in uncomfortable silence until Francis returned.

"How do you explain the fact that the four which Rubenstein claimed were positive never had any clinical symptoms, tested negative on three definitive tests, and all their mothers, fathers, aunts, sisters, and other relations were negative?" I asked Taylor as he changed his shoes.

"What do you mean 'no clinical symptoms'?" Taylor asked, as he pulled on a boot. "There were symptoms from the other farm," he insisted, leaning over to wipe the boots in the bucket Wayne had beside the truck.

"No, there were not," I said. "There were no symptoms in any of the animals, and none of our animals ever tested positive." Taylor looked at me, unsure whether to believe me.

Francis saw what was happening and joined in. "Why did we have to take the plastic off the roof of the solar barn? Were you afraid our sheep were floating and might have touched it?"

Wayne and Bill stayed quiet.

After they inspected the skeleton of the barn, Race asked to see our compost pile. Bill and I walked in the front followed by Davis, Larry, and Francis, while Race and Taylor lagged behind. From the compost pile you could see Freddie the llama and the sheep in a paddock in the adjacent field.

Freddie saw Bill and screeched. Immediately he rounded up the sheep and geese and chased them into the side of the paddock farthest from us. The geese were not used to being herded and balked, but Freddie jumped in the air, his hooves striking the ground close to them, until he finally had them all in the corner. Once that was accomplished he ran to the side of the fence nearest us and stared directly at Bill, his ears flattened against his head. I turned and pointed him out to Davis, "Looks like someone remembers Bill," I said. Freddie looked like he was ready to pulverize him. Secretly I was amazed. Freddie remembered Bill from the seizure and saw him as a predator. After inspecting the compost pile, Bill assured Taylor and Race it was not necessary to walk the field where Freddie was.

I wasn't the only one to find my spirit again. On September 22 there was a gathering of all fifty state commissioners of agriculture in Burlington, because Vermont Agriculture Commissioner Leon Graves was the president of the society. About a hundred farmers and protestors had gathered for a rally outside the conference, and there Heather gave the following speech entitled "Where Was Leon?"

> Hi! My name is Heather Faillace, and I am fifteen years old. Most of you have probably heard about my family, because on March 23 of this year the USDA stole our sheep. Today, the agricultural commissioners from fifty states are meeting here in Burlington. Why do we have state commissioners? To support and protect the family farms in each state. Leon Graves, our commissioner of agriculture, is now the president of the state commissioners' society.
>
> I would like to know where Leon has been in the past three years. Where was Leon when the USDA first came to our farm in 1998 and tried to force us to sell our sheep?
>
> Where was Leon when evidence was presented by European and American scientists showing our sheep were healthy?
>
> Where was Leon when my sister, brother, and I went out to our barn in a morning snowstorm and had twenty-seven armed federal agents kick us out of the barn and tell us we could not feed our sheep for the last time? How is allowing the federal government to come in and destroy an innocent family farm supporting agriculture? This man is supposed to be a model for the rest of the states to look up to.
>
> Where were you, Leon?
>
> Leon is out there supporting factory farms. Leon is allowing the use of genetically modified organisms (GMOs) in the state

of Vermont even though the majority of people want to see their use restricted. Leon is helping to destroy small family farms and sitting back and watching as federal officials come into our state and steal our property.

My sheep were stolen six months ago. Where are the test results? Where is the financial compensation? Our family was making our living farming, and now we are struggling to make ends meet. Is this what we want Leon to teach the rest of the country?

It's time for us small farmers to stand up and be heard. Do not allow the government to take control of our lives, and, most importantly, our food supply.

How can we do that? By supporting our local small farms, our family farms. I now manage our country store in East Warren, and I try to support as many local and Vermont farms and businesses as I can by selling their products. My father and sister make cheese on our farm and give cheesemaking courses. My family and I will continue farming. We will not be defeated by idiots in bureaucracy.

We will be living in East Warren, farming and working at educating farmers and consumers on the importance of local, organic food. We will be supporting agriculture and family farms.

Where will you be, Leon?

26

THE EMPRESS HAS NO CLOTHES

Detwiler's world began to unravel on October 18, 2001, when it was announced that a four-year British study investigating whether sheep initially diagnosed with scrapie were actually masking BSE was found to be worthless. In 1997 the British government had funded more than $325,000 to the Institute for Animal Health (IAH) to test sheep for BSE. From October 1990 to 1992 scrapie-infected brains from 2,867 sheep exhibiting clinical symptoms were collected at Veterinary Investigation Centres in the United Kingdom. By 1998 preliminary data indicated that sheep which were thought to have scrapie did, in fact, have BSE. However, only a few select scientists were privy to these preliminary results—and Detwiler was one of the few.

Was this why she urged the Vermont Department of Agriculture to put us under quarantine in July 1998?

When the study concluded in 2001, an independent audit of the laboratory did DNA testing on the brain samples and discovered 100 percent of the tissue was of *cow* origin, not *sheep*!

"It beggars belief," a British official commented, "that it took these scientists four years to work out that the sheep brains they were investigating for BSE were in fact cows' brains." Further investigation revealed that the original samples were improperly marked and contaminated, and the entire study was declared null and void. The findings of the audit were announced only two days before IAH was to present the results to the SEAC, which was prepared to recommend an immediate and massive slaughter of the UK sheep population. The UK Rural Affairs minister was quoted as saying, "We would have been much more embarrassed if we'd taken some very drastic action in relation to sheep, on the basis of those flawed tests."

But the USDA did take drastic action against our sheep. Our sheep were the only sheep in the world killed under suspicion of mad cow disease. It seemed probable that this flawed study was how Detwiler and the USDA were able to have the Vermont Congressional Delegation agree to the seizing of our sheep.

Davis said it best in his letter to Alfonso Torres, Deputy Administrator of USDA/APHIS.

> Dear Mr. Torres:
>
> The Emperor has no clothes. The recently disclosed "species barrier" debacle in Europe, described by Spongiform Encephalopathy Advisory Committee (SEAC) chairman Peter Smith as a "dreadful mistake," has finally provided the key to decoding the USDA's incoherent set of explanations for its Declaration of Extraordinary Emergency, quarantine, and destruction of my clients' business assets, livelihood, and, in some respects, their credibility.
>
> Dr. Detwiler often cited "information that she could not divulge" when asked why the USDA refused to give weight to the numerous negative test results from the Vermont sheep or to explain how Dr. Rubenstein's results, which admittedly were not "atypical TSE," might support the former Secretary's declaration. Clearly, she was trading on inside knowledge of erroneous preliminary data from the Institute of Animal Health study to justify and draw political support for USDA's conduct.
>
> It is unsound, unprincipled, and unethical for any research scientist to act on the basis of preliminary data that has not and cannot be verified and validated. The reasons for this are obvious. Such as here, where scientists were looking at tissue from the wrong species, it is simply too risky to act before sound conclusions, grounded in fact, can be reached. To the extent the proper parties must now bear the burden of assuming the risk, let it be perfectly clear that those parties are Dr. Detwiler, APHIS, and the USDA.
>
> Dr. Detwiler has already said that the USDA has histopathology and immunohistochemical assay test results from the Vermont sheep. Though still kept secret, they are almost certainly unremarkable. It is time to put this matter to an end.

Please confirm that Ag-Innovations may now make full and unencumbered use of its pastures, equipment, and lands, that no site remediation is necessary, and that the Declaration of Emergency is no longer in effect. In addition, please expedite payment as outlined in the Property Valuation materials we submitted to your office on August 16th.

The potential consequences of the erroneous data would have been devastating to the sheep industry worldwide and caused unnecessary panic in consumers. How did this fiasco happen? Money. Plain and simple—money. There is a tremendous amount of government and private funding shoveled into the fear-mongering of TSEs. The UK's sister organization to the USDA, the Department for Environment, Food, and Rural Affairs (Defra), has given almost $30 million toward researching BSE in sheep. For a scientist, being the first to discover BSE in sheep would put one at the top of the funding list. And searching for BSE in sheep is not the only way scientists make money; development of diagnostic tests worth potentially millions if not billions of dollars is the next "gold rush" of the science world. Rubenstein and Gambetti recently received a patent for their "rapid prion-detection assay," which they helped develop for Genesis Bioventures, a multimillion-dollar company. If this assay was approved for testing American cattle and the USDA followed Europe's lead and tested all cattle over the age of thirty months, at approximately $20 per test and more than thirty-five million cattle per year to be tested, Genesis Bioventures would gross more than $700 million per year on cattle testing alone, with exponentially more if deer, elk, sheep, and humans were included in the testing.

On the one hand are the scientists who stand to receive funding, prestige, and possibly incredible wealth from the discovery of new diseases, the diagnostic tests for these diseases, and the possible treatments or cures. Most of the money is dispersed not to diseases of livestock, but to diseases that could affect humans. If a human health scare exists or is created, scientists stand to gain.

On the other hand are food producers and sellers who want to assure their consumers that their products are safe. The U.S. beef industry itself was worth more than $78 billion in 2005. The beef industry had a vested interest in taking the spotlight off cattle and transferring it to sheep as the potential cause or vectors of BSE.

In the middle of all this is the USDA. They are responsible for animal health and also for promotion of animal products, interests often in conflict. So in 1998 they paid the Harvard Center for Risk Analysis to "evaluate the robustness of U.S. measures to prevent the spread of BSE to animals and humans if it were to arise in this country." Three years and $500,000 later, Harvard determined the "U.S. is highly resistant to any introduction of BSE or a similar disease." The USDA boasted of the safety of American beef to our export markets, not because of extensive, vigilant testing of cattle, but because Harvard said so.

An interesting quote from the November 2001 Harvard Center for Risk Analysis report: "Currently there is no evidence that sheep and goats can develop the disease [BSE] after exposure to feed supplemented with the contaminated protein."

Our emotions were conflicting—we were vindicated by these new findings, yet our sheep were dead. There was nothing we could do to bring them back. The USDA tried their best to put us out of business: Despite repeated letters from Davis requesting its removal, Jan Carney still had the warning against our cheese on the Vermont Department of Health website, there were still no test results, we did not have permission to use our land, and the USDA refused to compensate us. But we would not be defeated. There had to be a reason we went through this. Something good would come out of it.

Francis's college application required him to write about a "significant day" in his life, and he wrote about the day of the sheep seizure and put together a collage of related articles. In December a letter in the mail was the beginning of the rainbow after a storm: Francis was accepted to Saint Lawrence University and received a nearly full scholarship!

In February, the Archers learned that they had to surrender the remaining original imported water buffalo. When they all tested negative, the Archers were allowed to keep the offspring. It was their best hoped-for result, because it allowed them to continue farming and not lose the genetics. The one-year anniversary of the seizure was coming soon, and Larry suggested we visit the Archers during the slow period for the store. The children were old enough to stay alone, and my parents promised to check in on them.

We flew to Seattle and took a few days to drive around the Olympic Peninsula, savoring our time together. The scenery was breathtaking, and we enjoyed many long hikes through the Olympic National Forest and the surrounding area. One afternoon I came across a pay phone at one of the rest areas for hikers and decided to call the children and check in. The phone was busy, so I called to check our messages. "You have seventeen messages," the recording said. I could feel my body physically react. Something was wrong. I hung up without checking the messages and called the children again. Jackie answered. "Where are you guys?!" she cried.

"What's wrong, Honey?" I asked.

"The USDA said two of our sheep tested positive."

I was silent. I knew there was no way our animals could have tested positive. What sort of game was the USDA up to now? Why did they announce it when we weren't home? Was it a coincidence?

"Mom, are you still there?" Jackie asked.

"I'm still here. How are you guys doing?"

"Francis is handling all the interviews. I don't want to talk to anyone," Jackie said quietly. "Why do these things always happen when you guys go away?"

I didn't have an answer.

I told Larry what happened, and he got on the phone to comfort Jackie. I walked away and stood near a picnic table, feelings of anger and helplessness surging through me. When would this end?

Once Larry had Jackie calmer, he phoned Davis and asked him to request copies of all the test results. According to the USDA website, the animals were tested with histopathology, IHC, and Western blot, and "the abnormal prion protein was detected by the Western blot test in all of the sheep that have tested positive for a TSE in these groups of animals." What did that mean? The quote didn't even make grammatical sense. Did it mean the animals were positive by histopathology, IHC, and Western blot? By only one test? Was there a confirmatory test?

I had more questions. Where were the test results on Mr. Freeman's sheep? Why were our results announced and not his?

Bobby Acord, administrator of USDA/APHIS, said, "These results confirm our previous conclusions were correct and that we took the appropriate preventative actions in confiscating these animals. USDA's actions to confiscate, sample, and destroy these sheep were on target. As a result of our vigilance, none of these confiscated animals entered the animal or human food supply."

When questioned about the delay in obtaining the results, Detwiler said, "Additional biosecurity measures were needed to be able to test the tissue from the sheep . . . and scientists had to make sure the testing procedures were satisfactory."

Our visit with the Archers remains a blur; most of our time was spent giving interviews over the phone and trying to get more details. The Archers took us to the Sooke Harbour Inn on Vancouver Island for an amazing meal and an overnight stay in one of their beautiful rooms with a fireplace, spa tub, a bottle of port, and a private balcony with views of the water and the Olympic Mountains. It should have been one of the most romantic evenings, but instead we lay awake in bed, thinking of our children, our sheep, and wishing we were home.

The real story came out in May when Larry and I received a large box full of *all* the test results.

Every single animal was negative by histopathology.

Every single animal was negative by IHC.

Every single animal was negative by Rubenstein's Western blot test using the same antibody he used to find the supposed "positives" on four of Mr. Freeman's sheep.

We were stunned. All our sheep were negative.

But Rubenstein did two additional tests using different antibodies and concluded, "As described in our protocol, all samples were analyzed in the absence and presence of proteinase-K (PK) digestion. PrPc [PrPsen] is completely digested following PK treatment while the PrPsc [PrPres] banding pattern exhibits a downward shift in molecular weight following PK treatment. Although all scrapie-infected control samples demonstrated this pattern after PK digestion, samples 4677 and 4703 did not appear to. Instead the samples were in the absence and presence of PK digestion banded similarly and appeared as if they were all PK-treated. This result can be explained if the samples were autolyzed and/or by high levels of proteolytic enzyme activity in the tissue. [What Rubenstein was saying here was that the two samples were degraded due to their long storage in the freezer, and therefore not suitable for testing.]

"At my request, the blots were also examined by a number of my colleagues who have also been involved in studying prion diseases for many years. They all concur with my findings."

Rubenstein's report was ambiguous. It could be interpreted that the animals tested positive, and it could be interpreted that because of the degradation of the samples, the results were negative.

When I called Bruno Oesch of Prionics and shared the test results with him, Bruno said that similar results have come from freezer-burned tissue and, in such a case, Prionics throws out the samples. But the USDA claimed they were evidence of positive results.

Why did the USDA insist on different standards for our sheep? What about the fact that if these animals were cattle (deer, elk, humans, or American sheep) they would have been declared negative by the histopathology and confirmed to be negative by the IHC? That would have been the end of testing.

In the midst of all this, the USDA sent Davis a check. What was the value they put on the sheep after we produced all the required voluminous documentation? Exactly what the USDA appraisers claimed was the value in July of 2000 ($215,005), and not a penny more. For months we had given the USDA everything they asked for, and Davis attended meetings with the USDA in Vermont and Washington, D.C. The USDA repeatedly said they were working on finalizing the figure, and yet it was all a lie. They dragged their feet on our compensation. The check arrived two days before the USDA announced that two of our sheep tested positive. The USDA wanted to tell the press that the sheep were positive and that we had been paid.

It was a drop in the bucket compared to what we owed and what our business was worth. The first thing to come out of the check was $42,000 for legal fees, next we paid off credit cards we had used to survive, and then we gave each of the children $1,000. It was a pittance for the amount they had worked, but we wanted to show our appreciation. What was left we saved and would use to survive until our cheese business and the store were successful again.

The USDA followed the compensation check and the announcement of the supposed positives with a letter informing us our farm, the llama, and the rams were once again under quarantine, but this time a *federal* quarantine for five years to "prevent the spread of an atypical, undifferentiated transmissible spongiform encephalopathy in the United States."

Let me get this straight. Our animals were seized and killed in March of 2001. The USDA collected tissues and tested the animals using IHC around June 10. All animals were negative. The last week in July they performed the histopathologies. Again all animals tested negative. When the tissue was nine months old and freezer burned, Rubenstein ran the test the USDA claimed in court was their "gold standard." Again, all sheep tested negative. Rubenstein fiddled with the test until he found something unusual in two samples (from ram lambs that were only eleven and twenty-one months old). The USDA waited four months and then publicly announced that they had found two positives.

First of all, why were so many tests run?

Who told Rubenstein "if at first you don't succeed in finding a positive, try, try again"? How could an eleven-month-old lamb be positive?

And how could two ram lambs be positive when all their mothers, fathers, sisters, and relatives were negative? The USDA offered no explanations and repeatedly told the press, "We stand by our results."

The USDA's vendetta against us not only included the five-year quarantine on the land, which prohibited us from having sheep, cows, or goats, but all of our equipment used on the sheep—from the hoof trimmers to the feeders, fences, and milking parlor—had to be destroyed. And just for good measure, all the hay from our farm would be purchased and incinerated.

Dr. Taylor wrote to Detwiler and said he liked her suggestion that either the top six inches of topsoil be removed from areas where there was manure or compost, or that multiple hypochlorite treatments of the surface soil take place.

Bill Smith hired a local waste company to bring large trash receptacles for Francis to fill with the equipment, hay, bedding, and manure. Francis was paid $20 an hour for his work, but the more he threw away the angrier he became. He watched his and his sisters' shepherd's crooks go into the dumpster, was forced to throw out his "sheep care" kit he won at one of the Wisconsin conferences, and gathered all thirty-one rolls of flexinet fences and added them to the pile. We rented a skid steer, and Francis cleaned out the barn and then proceeded to dismantle the same building he lovingly built three years earlier. All this for a disease that did not exist.

The USDA announced the cleaning and disinfection of our farm with great pride. They would show the world how conservative they were at keeping mad cow disease out of this country. Dumpster after dumpster came and went, off to the incinerator in Massachusetts, or so the USDA claimed.

Larry had an inkling that the entire thing was a ruse and decided to trail the truck that collected the dumpster. The truck went nineteen miles north to the Moretown dump. Not to Massachusetts. Not to an incinerator, but to a local landfill.

As Larry took pictures of the dumping, he decided it was the final straw. It was time to sue the USDA for fraud.

We filed a motion in district court challenging the declaration of extraordinary emergency. Davis argued the case should not have been declared moot, the USDA's representations to the second circuit court were false and misleading, and the USDA took actions that continued to cause us to suffer economic and other harm. Davis asked for discovery. Judge Murtha quickly denied the motion that then allowed us to appeal to the second circuit court.

We were not the only ones to sue the USDA. Dr. Mary Jo Schmerr took the USDA to court citing gender discrimination. While she worked on developing her blood test at the USDA's Agricultural Research Service Laboratory in Ames, Iowa, she encountered a range of problems with her superiors: She was treated badly when she complained about the treatment of a female researcher from the Czech Republic; when she questioned safety at the lab; and when she tried, unsuccessfully, to get permission to accept speaking invitations. The breaking point came in early 2000 when the USDA sent Dr. Schmerr sheep samples for evaluation with her blood test. The results would prove the worth of her procedure, but Dr. Schmerr discovered the samples had been tampered with. She argued in her case that the USDA was undermining the credibility and reputation of one of their own researchers.

In December 2002 a federal jury awarded Dr. Schmerr $1.3 million.

We refused to sit back and give up. We had a life to live and to enjoy. Larry and I gave speeches at local schools and colleges; Jackie received more and more recognition for her cheesemaking including a full page article with photo in *Gourmet* magazine; Heather took up cross-country skiing, cross-country running, and track all while helping to manage the store. She also established an extremely successful booth selling cheese at the local farmers' market.

Francis was very well received when he spoke to a local group whose mission was to defend property owners' rights against government regulation. Francis had also become so well-known for his passion for human rights that the Harvard University Carr Center for Human Rights Policy accepted him as an intern during the summer. He was the only high school student selected. After Harvard, Francis went to Saint Lawrence University where he studied government and business. His freshman year flew by, and before we knew it he was home for the summer.

The first week or two Francis was unusually tense. Finally he started talking about how coming home brought back all the memories and reawakened his unresolved anger. It hurt him to see the now empty fields where he once pasture-managed the sheep; to see the skeleton of the barn and the trees with "USDA Lies" still spray-painted on them.

In early May 2003, Francis told us he spotted a USDA vehicle going back and forth past the farm. I thought he was just being paranoid and on this particular morning I took his USDA sighting with a grain of salt. An hour or so later, though, Larry and a friend saw one of the USDA vehicles. That same afternoon it was announced on the radio that Canada had discovered their first indigenous case of BSE. Immediately the USDA closed the borders and stopped importation of all ruminants and ruminant products. The Canadian beef industry went into a tailspin, the Canadian dollar weakened, and share prices dropped for stock in McDonalds, Wendy's, Outback Steak House, and other industries closely connected to beef. Canadian beef was a $5 billion industry with more than 90 percent of that in exports to the United States.

The Canadian bovine that tested positive was suspected of having pneumonia and was slaughtered in January. (Why did it take until May to have test results?) Steve Kerr, Vermont's new commissioner of agriculture, told reporters that the Canadian/U.S. borders were closed and the ag department was trying to help farmers who planned on importing Canadian cattle to find other sources. He also said the department was trying to trace imported Canadian beef. So if the U.S. was at an increased risk of BSE, why was the USDA driving around our farm?

The answer arrived the following day when Todd Johnson called and asked to come out to the farm to inspect that all cleaning procedures had been completed. Wayne Zeilenga had already done this last summer and had "signed off" that everything was done to their satisfaction. Why was another inspection necessary nine months later? Before agreeing to the inspection I asked Todd. "So why were you sneaking around here yesterday?"

"Yesterday?" he asked, trying to buy some time.

"Yes, Francis and Larry saw you yesterday."

Todd stammered and made an excuse about "checking on things" and then hung up.

The next day, Todd arrived a few minutes after Davis. "I thought you had resigned from your job as state veterinarian?" Larry questioned him. A recent article in a Vermont paper explained how Todd resigned because the department was understaffed and underfunded. He was overworked and unable to handle all the responsibilities.

"I did. When Wayne left, I took over his position," Todd replied.

"Where did Wayne go?"

Todd sheepishly paused for a moment and said, "He's now working with Dr. Smith in Massachusetts."

So that's what happened, I thought. They all got promotions for killing our sheep, and Detwiler gets an award and accolades in respected scientific magazines.

"You must be really busy dealing with the case of BSE in Canada," Larry said.

"Yeah."

"Are you tracing all the cattle that were imported into Vermont?"

"No, we know that cattle came in from Quebec and Ontario, but we don't know of any that came in from Alberta [where the case of BSE was discovered]."

"Are you going to trace animals?"

"No. We don't think it is necessary," Todd replied nonchalantly.

Not necessary?! I felt like screaming! The USDA found it necessary to kill animals over a disease that did not exist in that species. Yet in cattle it was "not necessary" to even try to trace possible infected animals. (Larry had once made the comment to me that the title of this book should be *Arbitrary and Capricious*, and never did it seem more accurate than at that moment.)

When we walked up to the barn, Todd found a piece of plastic hanging from the frame and proceeded to pull it down.

"Are you afraid the sheep could have been on the roof?" Davis asked facetiously.

"Or maybe their breath touched the plastic," Larry piped up.

Todd ignored the comments and spotted a large tarp that had been used to cover machinery during the past winter. "What about this?" he asked.

Larry explained that it had nothing to do with the sheep and what its use was. "You can take it. We were going to throw it out because it was ripped."

"Sure, I'll take it," Todd replied, almost excitedly. He continued walking around the barn, gathering bits of trash in a black trash bag and a few pieces of plastic and wood that he loaded into his USDA-issued SUV. The large tarp was too big, so he drove back to Montpelier to get a van.

That evening as we had dinner the children listened quietly as we told them about Todd's strange visit. Finally Heather couldn't take it anymore and exclaimed, "Doesn't this make you angry?! Here Canada has a case of BSE, and where is the USDA vet for Vermont? He's not at the border; he's collecting bits of trash on our farm!"

She was right. What was the USDA's strange obsession with a small sheep farm in Vermont? For more than a year, Larry had encouraged me to write a book, but the wounds were too open. Now it was time to fight back. I began writing, and the words flowed with ease when I talked about England and how we built our business. Larry was happy to see me being productive and offered to run the store during the winters so I could write. The more I wrote, the more I felt like I was rediscovering myself, awakening the optimist within. There must have been magic in the air, because in December another phoenix rose—Heather was accepted to Middlebury College!

We were told by Middlebury's admissions staff that they actively sought students who were unique individuals. Heather may not have had the highest scores in her class, but her grades were good and she more than made up for it in creativity, kindness, tenacity, and confidence. She and I cried tears of joy as we read her welcoming letter. Heather's entrance essay was amazing, and she included her "Where Was Leon?" speech along with articles about the sheep situation—another reason to be thankful for our sheep.

Two more miracles happened on a gorgeous day in spring as the sunshine streamed through the windows of the store. Early afternoons this time of year were generally quiet at the store, and I used the time to organize products on the shelves while Larry was in the cheese facility making cheese. I jumped when Terry, our mailman, came through the door of the store with a box of cheese and a large pile of mail. Sitting on top was a letter from Middlebury College.

"This might be the piece of mail you've been looking for," he said, pointing to the envelope.

I had applied for all possible financial aid and was waiting to see how much Heather would receive. We were thrilled over her acceptance to Middlebury but concerned about the $42,000 annual tuition. I put the box of cheese in the cooler and went to open the letter with Larry.

An hour later a couple came into the store, just as Heather called from her track meet. "Did we find out yet?" She asked breathlessly over the phone. "Did we get anything in the mail?"

"I think you'd better sit down, Heather," I told her.

The woman shopping in the store moved closer, curious about my conversation.

"The letter from Middlebury came," I said.

"What did it say?" Heather asked anxiously, not wanting to wait any longer.

"You got a *full* scholarship to Middlebury!" I exclaimed.

Heather sobbed on the other end of the phone, and the woman standing near me was now crying, too. She came over and gave me a big hug. "I don't even know you, and I am so happy for you!" she said between the tears. I hugged her back. Larry came in the store to see the woman and me wiping tears from our eyes. I learned that Robin and her husband, Ray, had just moved to the valley from Connecticut and worked together out of their home. They shared our passion for food and loved having our store nearby. That day was the beginning of a beautiful friendship for me and a very exciting chapter for Heather. A year later Jackie was also accepted to Middlebury—with another full scholarship!

27

BLAME CANADA (AND THE MEDIA)

The inevitable happened in December 2003 when the first case of BSE was discovered in the United States. On December 9 a cow that had complications giving birth was sent to a slaughterhouse in Moses Lake, Washington. Because of her inability to stand, she was labeled as a "downer." She had no neurological symptoms. The USDA, in its infinite wisdom, only tests a very small number of slaughtered cows, and then only cattle from the "downer" population. The cow was slaughtered and the meat inspector stamped the carcass "tested and passed" and shipped it into the human and animal food chains *before* the brain samples were actually tested. The samples were sent to NVSL, but because there were no clinical symptoms, the brain joined the regular testing pile. On December 22 the brain tissue was finally tested using histopathology and IHC. Both tests were positive. A USDA employee hand-carried a sample to Weybridge, England, and on December 23, the United States had their first case of confirmed BSE.

Where was Detwiler? Gone. She had resigned from the USDA in August, just in time to duck the first case of "mad cow" disease in the United States.

The USDA's spin doctors now had to work overtime to control the fallout. Within hours of the confirmed positive, fifty-three countries banned U.S. beef and beef products and almost $3 billion of the U.S. export market disappeared. Live cattle markets in the United States dropped 20 percent almost overnight. Two days after Weybridge confirmed the positive, the USDA sent a team from Washington to Japan to pursue trade talks. Eight days later another group from the USDA flew to Mexico to attempt to convince our second-largest importer to resume trade.

For the American consumer, the USDA and the National Cattlemen's Beef Association (NCBA) had their lines well rehearsed and wasted no time being interviewed and repeating them ad nauseam: "The U.S. beef supply is the safest in the world. Consumers shouldn't be afraid to eat meat because the infected material from the cow wouldn't enter the food supply. The discovery of this case actually shows that the government surveillance system works."

The problem was the system was *not* working. The infected cow was turned into ground beef, combined with the meat from hundreds of other cattle, and shipped to Washington, Oregon, Alaska, Montana, Hawaii, Idaho, Nevada, California, and as far away as Guam. More than 4.7 tons of meat were recalled. But by the time the recall was finally enacted, most of the meat had already been consumed. Despite this, the USDA assured consumers the infected meat was safe.

Consumers were *not* safe. Especially not in California where the state was forbidden by law to reveal where the recalled meat was sold. The USDA attempted to control the panic and said the meat posed no health risk and the recall was only precautionary.

The USDA also struggled to maintain a cover-up. First there was the discrepancy of the age of the positive animal. The farm manager said the cow was four and a half years old, but the USDA insisted it had to be six and a half years old (to make it born before the United States implemented the meat and bonemeal feed ban). Yet when her offspring were traced, it was discovered that the infected cow's first calf was born in 2001. A cow's first birth experience is usually around two years of age, which would make this particular cow a little over four years old, not six and a half as the USDA wanted her to be. The cow came from a very large dairy farm, and, in order to maintain profitability, cows that were barren for two years would be culled.

When BSE is found only in imported cattle, some countries will allow the export market to be reopened. So where did the USDA claim this individual cow came from? Canada. Blame Canada. Canada already had a case of BSE in an indigenous cow, what harm would there be in adding another one? The United States has a long history of bullying Canada, and the beef-ruled USDA was as big a ruffian as any. When Canada announced its first case of BSE, the U.S. immediately shut the door on all Canadian exports of ruminants and ruminant products. This crippled the $5 billion Canadian beef export industry as more than 90 percent of their beef product exports and 99.6 percent of live cattle exports were to the United States, as were

$20 million worth of sheep and sheep products. Canada agreed to say the Washington cow was theirs. (Once the two countries agreed the cow originated in Canada, Canada received some very lucrative contracts in the restoration of Iraq.)

The USDA claimed the cow was a downer, but three eyewitnesses, including the trucker and slaughterhouse worker who killed the animal, said she was not. The three men accused the USDA of forging documents. A FOIA by the United Press revealed the positive animal was not tested for antibiotics nor had a temperature reading been taken—both requirements for downer cattle. When asked if the USDA ordered the documents to be forged, the USDA spokeswoman repeated five times, "I cannot fathom that would happen." She would not put blame directly on Secretary Ann Veneman, but when queried if some other high-ranking USDA official could have requested forged documents, the spokeswoman repeated, "I cannot fathom that would happen."

The USDA wanted to make sure no more BSE cases were uncovered and immediately told the slaughterhouse that discovered the infected cow to *stop* taking samples for BSE testing. Common sense would dictate that once a positive was discovered, the USDA should *increase* their testing program. In 2003 the USDA tested 20,543 cattle for BSE, and in 2004 (the year immediately following the discovery of the positive BSE case) the USDA boasted they would test more than 200,000 in a one-time effort to determine the level of BSE. By December of 2004 only 15,513 were sampled—a *reduction* of more than 32 percent in testing.

Japan tested every single animal sent to slaughter and wanted the United States to do the same. The USDA refused. Creekstone Farms, a Kansas company that raised and slaughtered all natural beef, had a lucrative export market with Japan and volunteered to test all their cattle for BSE. The USDA threatened Creekstone with a lawsuit and possible imprisonment if Creekstone established their own self-funded testing program. The USDA's reason was fear that "private laboratories would report false positives." But false positives on rapid tests are rare, and as the USDA's own protocol dictates, follow-up confirmatory tests can prevent unnecessary scares to the market. After all, only the USDA would intentionally obtain false positives and create an unwarranted health scare.

Another USDA reason for refusing Creekstone the authorization to perform their own testing was "there is no scientific justification for 100 percent testing because the disease does not appear in younger animals." Yet the USDA alleged to have made the unprecedented finding of an "atypical TSE of foreign origin" in Mr. Freeman's thirteen-month-old lambs.

What a different tune the USDA and NCBA were singing in this case versus the fear mongering they created with our sheep. The list of inconsistencies grew longer every day. The infected cow was from a 4,000-head dairy farm owned by a large animal veterinarian. Only 131 out of 4,000 cattle were killed and tested. Our entire flock of 125 sheep was killed.

The Washington farm was allowed to continue milking and shipping milk. Jan Carney kept the warning against our cheese on her website for almost four years (three years after our sheep were destroyed) and would only remove it when we finally threatened to sue her.

The USDA quarantined the Washington farm for one month. Our farm was quarantined for five years.

Less than two months after the identification of the positive case in Washington and before all the herdmates and possible birth cohorts could be discovered, the USDA declared it was an "isolated case" and shut down the ongoing investigation.

The United States announced the first case of BSE in an indigenous cow in June 2005. What was most shocking about this? Seven months earlier, scientists from Texas A&M University had found the same animal to be positive for BSE. Tissue samples were then sent to NVSL where additional tests yielded positive results. When all was said and done, six tests came back positive, while one test was negative. Yet at the time, the USDA declared the animal negative—free of BSE.

Scientists at Ames urged the USDA officials in Washington to perform additional tests, but they refused. Their reasoning? "Conducting additional tests would undermine confidence in the USDA's testing protocols." Sounds familiar.

But when Inspector General Phyllis Fong, the head of the watchdog agency within the USDA, discovered "voluminous records" that indicated conflicting test results, she ordered additional tests. Samples were then sent to Weybridge, England, and confirmed to be positive. The USDA waited until late on a Friday evening to announce the results via a teleconference with Ag Secretary Mike Johanns (who replaced Ann Veneman when she resigned in November 2004, only hours after the animal first tested positive) and Dr. John Clifford, the chief veterinary officer of APHIS.

Dr. Clifford opened the call, "Good evening everyone, and thank you for joining us late on a Friday evening. I certainly appreciate your getting on with us on such short notice for an update of our BSE surveillance.

"The IHC is an internationally recognized confirmatory test for BSE. They [NVSL] had run the IHC flawlessly, and we're confident in every result that's resulted from that IHC. In order to find a positive in this particular case with this Western blot, they had to enhance or enrich it. So they had to use twenty times the amount. You would have to use about twenty times the amount of tissue for this to determine to be a positive or reactive on the Western blot."

After describing the Western blot procedure Dr. Clifford said, "Basically that [protein] is run through a gel-type separation using specific antibodies that will give you those bands. And they look at those bands and the molecular weight of those bands to determine the outcome of that test.

"We would never make a decision about changing protocol in a knee-jerk sort of way. We would certainly want to debate that. We would want to get a lot of good scientific analysis. So it's not something that we would do just very, very quickly."

Again the inconsistencies between treatment of cattle and treatment of our sheep were glaringly obvious: All our sheep were negative by IHC performed at NVSL and by Katharine O'Rourke. Rubenstein didn't use molecular weight markers. And Rubenstein didn't hesitate to change the protocol when testing our sheep. He used *three* different antibodies until he achieved his desired result.

This time the USDA actually managed to upset the National Cattlemen's Beef Association. The NCBA was publicly outraged and demanded that the BSE testing program "must operate under a consistent and established testing protocol. NCBA remains committed to a science-based approach in addressing these concerns, but we simply cannot tolerate actions that serve political pressures or pseudo-science over a sound surveillance program."

When the beef markets fell even further, who did the USDA blame? The media. Secretary Johanns complained that the news coverage was inaccurate and had cost the beef industry billions of dollars. Johanns said the threat to humans from BSE was "minuscule" and had been overblown in the media. Arbitrary and capricious, once again.

What do cheese, wine, bananas, beef, and airplanes have in common? They all have been pawns in sanctions and trade wars between countries. For example, in 1988 the EU refused to accept hormone-laden American beef and genetically modified grain due to the fact that hormones and GMOs are banned in the EU. The United States retaliated by slamming more than $300 million of tariffs on European imports such as Roquefort cheese from France, Belgian chicory (endive), pecorino cheese from Italy, cashmere from Scotland, and the list went on. The EU decided to try and even the score by preventing American bananas from entering the EU market.

When the tariffs began to take their toll on countries within the EU, a truce was reached and the United States was allowed to resume shipping beef, provided it was hormone free. Shortly after the ban was lifted in 1999, EU scientists tested five hundred randomly sampled imported U.S. beef products labeled as hormone-free, and sixty tested positive for hormones, including the hormone diethyl-stilbestrol, a known carcinogen.

The positive case of BSE in 2005 hurt the injured U.S. beef export market even more. But the USDA was up to their usual bullying tactics. According to an article in the *New York Times*, "Mr. Johanns' undersecretary for farm and foreign agriculture services, J. B. Penn, worked at a consulting firm serving the [beef] industry. Last week according to the Kyodo news service in Japan, a group of Japanese lawmakers who visited Mr. Penn in Washington accused him of 'threatening' them with trade retaliation and saying that the United States' patience was growing short and that they should simply accept American beef."

In December 2005, two years after the discovery of the first case of BSE in the United States, Japan reopened the door for U.S. beef with the agreement that only products from animals less than twenty months old were allowed and no specified risk materials (SRM), such as brain and spinal cord, were allowed. One month later a shipment of U.S. beef stamped "USDA Approved" arrived in Japan and contained SRM. The shipment was traced to a veal plant in Ohio. Mysteriously, only one week after disclosure of the error, the 40,000-square-foot-plant burned to the ground. A week later Japan reinstituted the ban on American beef.

The United States was not alone in employing intimidation tactics. Twenty-four hours after Canada was refused a WTO review of Brazil's aircraft subsidy program, Canada banned Brazilian beef and recalled all Brazilian beef products from Canadian shelves for "fear of mad cow disease"—despite the fact that Brazil never had a case of BSE. Canadian Food Inspection Agency officials claimed the action was "to ensure the security and integrity of Canada's food safety system." But food safety had nothing to do with corporate interests. Brazilian beef was safe. Food was manipulated in the global marketplace without regard to the panic and fear the actions would have on the general public.

Our second circuit court case was reopened in April 2004. Davis explained the process to Larry and me and the possible outcomes: The three appeals court judges could decide the case should not have been declared moot and then remand it back to district court (Judge Murtha) or they could say the USDA was correct and the case could not be reopened. Either way we figured we had nothing to lose.

It was cloudy, gray, and drizzly when Larry, Heather, Jackie, my sister Becky, and I arrived at Foley Square in New York City, but we were too excited to really notice. At 9:30 we went to the courthouse, through security, and rode the elevator to the seventeenth floor where Davis was waiting for us in the lobby. We quickly hugged one another, and then Davis said, "I've decided to go in strong, to show how the USDA lied, and not pull any punches." He explained his tactics, and we wholeheartedly agreed.

Davis led us into the beautiful courtroom with thirty-foot-high wooden ceilings covered in ornamental brown and gold raised designs. A long, narrow table with six leather chairs and a podium were flanked by two large television screens. This enabled the audience to see the face of the lawyer as he or she spoke with the judges. Three judges each had their own large, leather chair on a raised platform to allow the judges a clear view of the entire room.

We sat in the front row of the audience area, while Davis plopped into a leather chair, grinning, enjoying the luxury. Soon the room filled with more lawyers, but there was no sign of a lawyer for the USDA. Davis thought the USDA might decide to use the video-conferencing option and not send a lawyer.

While we waited, Davis pulled out a few pages from his briefcase. "I finally found the statute that the USDA used to quarantine the land. Would you believe that it was the last page of the last chapter and is listed under the 'Tree Assistance Program'?"

When he handed the pages to Larry and me, I noticed that the date the statute was written—May 2, 2002—almost a month after the USDA announced the supposed two positives from our flock. The statute was written expressly for us. As I handed the pages back to Davis, a tall man with a dark gray suit and light gray hair and beard walked into the room. He was David Kirby, lawyer for the USDA, and he smiled confidently. "Just like the rest of the USDA lawyers," I thought. "They have nothing to lose. Sheer size and power are on their side, and they are accustomed to winning."

The order for the morning cases was announced, and ours was the fourth. The first case involved a Chinese woman fighting the INS against getting deported back to China. The next case involved an Indian man fighting a computer company for being fired. I was amazed the third case had even made its way into second circuit court. A man had driven over the Vermont border at Highgate Center into Canada with $66,000 dollars in his vehicle and claimed that his crossing the border was accidental and he did not see the American border patrol building. The case didn't stand a snowball's chance. After listening to the lawyer for the plaintiff plead the case, the judge in the middle responded, "A wink and a nod mean nothing to a blind horse."

Now it was Davis's turn. The judges normally allotted five minutes to each lawyer, but for our case they gave ten minutes to each side. Davis stood at the podium in his light gray suit with yellow Polo tie and introduced himself, ready to deliver his ten-minute speech. No sooner had he said two sentences when the judge to the left interrupted him and asked, "If the government lied and was deceitful, why not fight the quarantine order?"

Davis began to explain how the case should not have been declared moot, when the judge in the middle cut him off. "What about the prior judgment is hampering you from suing the USDA regarding the quarantine?" he asked.

Again Davis began describing the steps that brought us back to second circuit court, and again he was unable to finish. The center judge explained that the declaration of emergency was not challenge-able and how the judges wiped out Murtha's judgment, which ended up being in our favor. "So what is stopping you from going back to district court?" the middle judge asked Davis again.

"Your Honor, the USDA lawyer, Paul Van de Graaf, stood in this very courtroom and said that once the sheep were seized there would be no further action. The argument that our time limit has been sur-passed is invalid due—"

"But what is stopping you?" the judge to the left asked, again.

"If I can explain," Davis said. "The USDA lied when they said—"

"If the mad government behaved as badly as it appears, you have a case. But why not sue in district court?" the judge in the center asked.

Things were beginning to get heated. Davis thought he would have ten minutes to explain his points, but the judges were stuck on this question. "Your Honor, my time has expired, and I will hand things over to Mr. Kirby." And with that he sat down.

David Kirby stood up, unable to suppress a grin. "Your Honor, I am David Kirby, counsel for the United States Department of Agriculture, et cetera.

"Your Honor," Kirby began, "the USDA—"

"Does anything stand in the way of this case going back to district court?" the judge in the center asked.

Kirby appeared taken aback. "Your Honor, if you remember—"

The judge on the right suddenly spoke up, "Can you answer the question yes or no?"

"According to the . . . " Kirby started.

"Look," the middle judge interrupted, "if it makes a difference you won't win. If you say 'yes' we will look at the case, if you say 'no' you win today, but the case goes below [back to district court]."

"Well, Your Honor, I would like to win today," Kirby replied smiling, again. The conversation then turned back to rule 60(b)6 and the USDA's claim that our time limit to reopen the case had expired. Davis explained how the USDA held on to the test results, waited months to release them, and followed the release of the results with the quarantine.

"What is the time limit on 60(b)6?" the middle judge asked Kirby. Kirby hemmed and hawed. When the question was repeated he finally quietly replied, "There is none, your honor."

"It appears the government intentionally wanted to get the decision mooted," the judge continued.

"You paint a picture of horrendous government action," Kirby replied.

"Well, disingenuous becomes fraud when it is the U.S. government we are talking about," the judge warned.

"I want to assure the court we are not being disingenuous," Kirby insisted.

Davis then stood and told the court that our accusations were not directed at the lawyers involved with the case, but the USDA officials. He thanked the judges.

On to the fifth case.

We quickly exited the room. I was confused. Davis was beaming. "The judges have basically given us a road map on how to sue the USDA," he told us once we were in the lobby. "I wasn't sure at first if we should try to reopen this case or just sue the USDA based on the quarantine. I'm glad we chose to go this route."

"So this means that the USDA can't stop us from suing them?" Jackie asked.

"That's right. They told the judges they won't put any obstacles in our way," Davis said with a smile.

28

FINAL PROOF

Inever realized how extensively government agencies used manipulation or how far their employees would actually go for power and prestige, until September 2005. In years to come I will look back at this date as the time I lost my layer of innocence. It was the last time I could claim to be blissfully naive.

Raised in a home filled with love and security, I was taught from a very young age that A+B=C, and if you always did what was right, all would be well. I had tremendous respect for people who had successfully mastered the corporate world, for scientists, for judges, and for government officials. These people, I believed, had everyone's best interests at heart. They had to do what was right. They were not to be questioned. What they said was to be accepted as the truth. But one small, silver, computer CD shattered my entire belief system.

Judge Murtha requested that the USDA supply the complete administrative record to our lawyer. We met with Davis, and he gave us a copy of the CD and an accompanying index.

"Check it out," he said. "It's probably just a rehash of everything they've already submitted."

The CD sat on my desk for a few days until I finally got around to opening it. On first glance it was a list of documents organized into a PowerPoint-style program. I flipped through page after page and saw that Davis was right, they were previous submissions. Nothing new.

But there were only a couple dozen documents, and the index was single-spaced and fourteen pages long. Where were the rest of the documents?

I began digging deeper and deeper into file folders on the disk and discovered more than two thousand pages of additional material, most of which I had seen before, but then I came across one with Detwiler's name and address on top and a subject line that read: "Re-evaluation of positive results of the Faillace flock in light of new information about subsequent BSL-3 closure."

The "BSL-3 closure" memo revealed that Rubenstein's lab had been shut down! The same laboratory that produced all the supposed positives on our flock and on Mr. Freeman's was closed due to "gross negligence," and this fact was buried in the record. There was no reference to the lab closure or subsequent investigation by the USDA anywhere in the entire index.

I grabbed the USDA letter that accompanied the CD and was shocked when I saw that the USDA admitted in the letter to Judge Murtha that Rubenstein's lab was shut down. According to author Michael Doerrer, special assistant to the Deputy Administrator of Veterinary Services at the USDA/APHIS, he could not locate a final decision for the administrator to impose the five-year quarantine on our farm, only a draft. "Thus the document may contain representations that are not legally accurate," he wrote.

Sure enough only a draft existed, and only one name was on it. One person wrote the draft and sent it to the administrator—Dr. Linda Detwiler.

For the next two weeks I spent day after day reading every single file on the CD and pieced together a story of lies, intrigue, and corruption.

Detwiler's connection to Rubenstein went back more than a decade. Rubenstein helped Detwiler gain recognition and establish credibility through joint publications in scientific journals, while Detwiler assured Rubenstein a steady flow of USDA funding for his projects. Rubenstein's lab supplied all the antibodies for TSE testing for the National Veterinary Services Laboratory in Ames, Iowa, and according to USDA documents, these projects were "directed by Dr. Linda Detwiler."

On July 7, 2000, the USDA approved Rubenstein's facility for biosafety level 3. Yet Rubenstein tested Mr. Freeman's sheep on June 22 and 28, 2000. This meant that Rubenstein's lab was not inspected or approved when he ran the tests on Mr. Freeman's sheep. According to the USDA, a laboratory must be a biosafety level 3 to test for BSE. Did Rubenstein receive his higher rating in exchange for the positive test results?

The USDA's administrative record also revealed that Rubenstein had lied under oath. When questioned by Melissa Ranaldo in July

2000 regarding articles he published about Western blots and sheep, Rubenstein referred to a USDA study where a group of laboratories across the United States were given blind, coded samples and asked to test for PrPres using various techniques. Rubenstein claimed 100 percent accuracy using his Western blot: "When the code was broken all the positive samples that were sent to me came out positive and all the negative samples that were sent to me came out negative."

Included on the CD was this very study, and of all the samples he received, Rubenstein had erroneously declared two of the negative samples positive and another negative sample suspect. (If you wanted a positive result on a negative sample, Rubenstein was your man.)

By the first week of August 2001, Detwiler had all the test results on our sheep and Mr. Freeman's. The National Veterinary Services Laboratory tested the animals using histopathology and IHC, and every sample was negative. But plans were already in place for Drs. Race and Taylor to inspect the farms three weeks later, so Detwiler kept the results secret, and the USDA spin doctors held the media at bay with a variety of creative excuses as to why there were no test results yet.

Two months later, when the UK announced the sheep study fiasco, Detwiler's credibility quickly diminished. She must have been mortified. She was supposed to save the United States from BSE, particularly BSE in sheep. She would be the first person to take measures against BSE-infected sheep, and once the study was released it would prove her actions to have been prudent. People would admire her for her foresight. But when the UK results were released, people began questioning Detwiler.

Dr. Taylor of Scotland sent Detwiler a letter of his recommendations for disinfection and possible quarantining of our farm. He recommended the most stringent measures be taken, but noted, "These comments are based upon my assumption that only *clinically suspect* animals were subjected to laboratory investigation. If this was not the case, please advise so that I can modify my opinion." Detwiler never informed him that none of the animals had any clinical symptoms and all had tested negative on the two tests.

To make matters worse for Detwiler, Davis sent the "emperor has no clothes" letter to Alfonso Torres. Detwiler would not be humiliated like that. Two weeks later ninety-nine samples from our sheep were sent to her friend, Rubenstein.

This time, Rubenstein would not take any chances of being embarrassed in court and refused to test the sheep until a "protocol" was developed. But the resulting protocol was a worthless piece of paper. No specific antibody was required and neither were negative and positive controls.

Who told Rubenstein to keep testing until he created a positive?

Rubenstein tested the animals in December 2001, but the USDA did not announce the two supposed positives until April 2002. Why?

Detwiler was back on top in April 2002 when our results were released. The USDA felt justified with their actions, scientists were impressed she was astute enough to know there was something wrong with the sheep, and politicians were happy they had chosen the winning team to support. But her reign lasted a little more than a year longer.

On June 5, 2003, Rubenstein's lab was shut down for "inadequate facility and equipment maintenance." Detwiler saw the writing on the wall. Our sheep were negative, the British study was invalid, and BSE in the United States was a time bomb waiting to go off. A month later, Detwiler resigned and neglected to mention to anyone that Rubenstein's lab had been shut down.

Rubenstein steadfastly refused everyone access to his laboratory. No one was allowed in without his permission. Not even safety inspectors. When Rubenstein received his revered BSL-3 security level in July 2000, his technician prepared the entire laboratory for an official USDA inspection. But no one showed.

Three years later, something happened. Someone complained, and the state of New York decided to investigate Rubenstein's BSL-3 lab. The findings were shocking. The laboratory had never been inspected during the entire time of its BSL-3 status. Mice ran freely throughout the lab, mouse droppings were everywhere, and mice were living in drawers and co-mingling with genetically altered transgenic mice. The inspectors were unsure whether the mice they saw were feral, transgenic, or perhaps crosses of the two. Bags of mouse bedding that were supposed to be incinerated were piled high throughout the lab.

But mice were not the only problem. The autoclave for sterilizing the equipment had not been used for at least a year, and there were signs of extensive contamination throughout the entire laboratory. The sonicator was operated without containment during the early stages of BSL-3 laboratory operation (around the time of testing Mr. Freeman's sheep). "Consequently, the lab environment may have contaminated material present."

Mouse cages were either labeled improperly or not at all. And Rubenstein had TSE material without the appropriate registration certificate.

After an inspection of the closed laboratory, a USDA-appointed review panel determined the work practices and the equipment's poor state of repair in the BSL-3 laboratory resulted in conditions that facilitated potential contamination in the work environment. They concluded that "the accuracy of the TSE work in the BSL-3 laboratory must therefore be questioned."

This left the USDA with a huge dilemma. All the alleged positives on both flocks of Vermont sheep came out of this laboratory. If the USDA had to admit the results were invalid, they would immediately lose the lawsuit. And to make matters worse, at the same time the review panel was inspecting Rubenstein's lab, the United States had its first case of BSE. The last thing the USDA wanted was to lose a court case, particularly one about BSE.

In a move that can only be described as inviting the fox back into the henhouse to assess what happened to all the hens *she* killed, the USDA rehired Detwiler to review Rubenstein's test results and to conclude they were still valid.

In her very thorough manner, Detwiler wrote pages of reports explaining Rubenstein's results were perfectly acceptable, despite a wealth of evidence to the contrary. For good measure, Detwiler asked Dr. Paul Brown of the National Institutes of Health to review Rubenstein's test results on our sheep. Brown carried significant clout with his position. Detwiler knew her friend wouldn't let her down. Brown noted that Rubenstein found the sheep to be negative using his "gold standard" polyclonal antibody. "Either because of previous experience with false negatives using this antibody, or because it was felt that the samples merited additional testing, a second round of blots was made," Brown wrote.

Brown also observed that the blots had a heavy background "smearing" and the "absence of downward molecular weight shift with proteinase-K, both consistent with autolysis of the sheep brain samples."

Referring to the BSL-3 laboratory closure, Brown reiterated, "Laboratory documentation, especially labels of mice cages, was inadequate to the point that animals could not be reliably matched to inoculated material. The accuracy of the TSE work in the BSL-3 laboratory must therefore be questioned." Brown recommended the USDA run DNA testing to confirm that the two samples which the USDA claimed tested positive (72 and 98) were truly sheep samples.

Detwiler quickly obliged. Included on the CD were the results. DNA tests showed that samples 56, 69, 70, 71, 90, 97, and 99 were all sheep samples. The two alleged positives, 72 and 98, were *not* among the samples that were DNA-tested.

A month later, Larry and I drank cappuccino at Three Mountain Café on a beautiful autumn morning. "You seem a million miles away," Larry said, taking a bite of his bagel sandwich.

I looked up from my croissant and into his beautiful brown eyes. I loved his eyes.

"Where are you?" he asked again.

"Do you really think she's so bad?" I asked.

"I take it you're talking about Detwiler."

"Yes," I said, leaning over to brush some crumbs off his beard.

Earlier that morning Barry Simpson stopped by the house to ask about up-to-date information on the transmissibility of TSEs for a term paper his daughter Sarah was writing. It had been a few weeks since we last spoke with him, and he was unaware of all the documents we received from the USDA through the court hearing.

Larry explained how Rubenstein's lab was shut down and the fact that Detwiler wrote the recommendation for the quarantine but never got approval.

"When Dr. David Taylor of Scotland inspected our farm and made his recommendations, he wrote Detwiler saying 'these comments are based upon my assumption that only clinically-suspect animals were subjected to laboratory investigation. If this was not the case, please advise so that I can modify my opinion,'" I interjected.

"And not only were none of our animals *ever* clinically suspect," I continued, "but Detwiler had the test results on August 3, 2001, from Ames, Iowa, proving all our animals and Mr. Freeman's were negative, and Drs. Taylor and Race didn't inspect our farms until August 27 and

28, 2001. So Detwiler knew all the animals were negative but with-held this information from the very people who were supposed to make 'scientific' recommendations regarding the future of our farms!"

I took a bite of my croissant, still mulling everything over. I was always looking for the good in people and for years I remained con-vinced that what happened to us was a result of misinformation and a series of mistakes.

"But I'm seeing a pattern here," I told Larry as he sipped his drink. "Detwiler uses people. She used Taylor to get him to recommend the five-year quarantine, she used Vanopdenbosch to claim the cases of BSE in Belgian cattle were feed-borne, and she used Paul Brown.

"Everything traces back to her. She was the one who had the crim-inal investigation of you. All the documents were sent to her, and when the USDA needed to produce the administrative record they hired her to come back and 'sort everything out.' I think she was the 'mastermind' behind all this."

Larry shook his head in disagreement. "No, she thought she was the kingpin, but she was just a pawn and now she's being hung out to dry. It was obvious to me all along."

"What do you mean she's just a pawn?" I argued.

"Follow the money. Always follow the money," he said. "Look at all the money that was spent on us. Detwiler was just a mid-level bureau-crat. They don't allow someone of her stature to spend all that money unless there's someone higher up pulling the strings, some bigger fish."

I thought about the fact that not a single document from the beef industry had ever shown up in the court papers. There were plenty of letters, e-mails, and other correspondence from the American Sheep Industry Association and the Vermont Sheep Breeders Association, but nothing from the beef lobby. The only trace of that side of the story were the letters Detwiler requested saying Japan and Korea were concerned over the situation with our sheep and the perception of the United States having BSE.

"So who are these 'bigger fish'?" I wondered. Discovery would give us the answer. The administrative record had answered many ques-tions, some we didn't even know to ask, but discovery would allow us to depose witnesses, delve even deeper into the workings of the USDA, find out who was behind it all, who was the "man behind the curtain." We waited for Judge Murtha to set the date for discovery.

We may have had a huge battle with an agency of the executive branch, but ultimately the judicial system had forced the USDA to disclose their secrets, their lies, their cover-ups. But at least we had the system of checks and balances. My belief was slowly being restored. If one branch of government was out of hand, another could rein it in. After eight years of being under siege, I could feel my optimism returning.

I walked to my office and reread the USDA letter included with the CD. The USDA had to admit defeat. Rubenstein's lab was closed, Detwiler was gone, and there was no legal basis for the quarantine. As painful as it was to uncover the corruption, I took comfort in the fact that we had proof—final proof that our sheep were free of any disease.

On February 24, 2006, one month before the lifting of the quarantine, Judge Murtha finally ruled on the legality of the quarantine—in favor of the USDA.

AFTERWORD

Mad sheep. They were all mad sheep. Not our animals, but those in the government—everyone from the elected officials to the civil servants, through the judicial branch up to the president of the United States. No one was willing to question one another. No one would question the misapplication of the law and the blatant disregard of federal regulations, not to mention the general lack of common sense, the exploitation of science and outright corruption. Don't step out of line. Sheep flock together for safety—safety in numbers. And with our situation the government flocked together and stood with their heads staring at the ground, refusing to listen to anything but what was said within the flock. They were all mad.

When Judge Garvin Murtha ruled that the USDA's five-year quarantine on our farm was legal, despite the USDA's confession that it was not, the remaining vestiges of my faith in the legal system were shattered. If all else fails, an American citizen is supposed to have the right to turn to the courts for assistance. Judges are supposed to be impartial, open-minded, intelligent, unwilling and unable to be influenced by outside forces. How did our judicial system become so closed to listening to the truth?

Judge Murtha had all the documents he needed to make his decision by the last week in November 2005. We assumed we had nothing to lose: If Murtha ruled in our favor, we won; if Murtha ruled against us, we could go to the federal appeals court. The USDA was afraid of the appeals court. Either way it would be a positive outcome. But Judge Murtha waited until three weeks before the quarantine would be lifted to make his ruling, leaving us with no chance to appeal.

So now the sheep are gone, our possibilities of having the court proclaim the USDA's actions illegal are gone, and we are still waiting to get proper compensation. If the government seizes property and refuses to pay fair market value, a citizen's only recourse is to go to the

U.S. government's own claims court in Washington. Not only does the citizen have the expense of traveling to Washington, but only certain lawyers are trained to litigate in claims court (and their rates begin at $250 per hour).

Mr. Freeman had tried every possible method to work amicably with the USDA, but he was unsuccessful. In December 2004, he asked Larry and me to join him in suing the USDA in claims court, and he generously offered to cover the legal fees.

For eighteen months we attempted to settle with the USDA out of court, but as of June 2006, the USDA has still refused to compensate us properly. They have not even paid us the same amount they paid all the shepherds who voluntarily surrendered their animals. We have no choice left except to go to trial in claims court, which the lawyers estimate will cost more than $250,000. How is a small family farm supposed to survive when the government illegally seizes their healthy animals, illegally quarantines their farm for five years, refuses to pay even the minimum "fair market value" of their animals, and then forces the farmers to fight the USDA in the government's own court?

The third recorded case of BSE in the United States was discovered on a farm in Alabama in March 2006. The infected animal had been unable to stand, was examined and treated by a veterinarian but did not respond to treatment. So it was euthanized, a sample of brain tissue was taken, and the carcass was buried on the farm in a communal burial pile with other cattle.

The United States and Korea have an agreement that the export market to Korea will remain open as long as all cases of BSE are found in animals born before April 1998. By the time the test results on the Alabama cow came back, the animal was severely decomposed. The veterinarian for the farm claimed the animal was ten years old. USDA officials attempted to exhume the particular animal's carcass from the others in the burial pile, and found a set of teeth that would substantiate their claim.

Despite the fact that the cow had a six-week-old calf still living at the farm, the USDA determined that no quarantine of the farm was necessary. But when Dr. Peter Fernandez, associate administrator of APHIS, reported the case of BSE to the World Organization for Animal Health (OIE), he said a quarantine *was* imposed on the farm.

Surprisingly, the international guidelines set by the OIE to determine a country's risk for BSE were recently changed. Previously there was a seven-year wait after the discovery of a case of BSE before a country could be considered in the "negligible risk" category. (A "negligible risk" rating, the lowest of three levels, improves a country's ability to negotiate export possibilities and requires fewer safety restrictions.) The new guidelines are based on the age of the last animal diagnosed with BSE. Now a country can be considered in the "negligible risk" category if more than eleven years from the date of birth of the infected animal has passed. According to an OIE chairman, Alex Theirmann, "This is much better for the United States." (Especially since the Alabama cow was more than ten years old.)

In addition to having worked for the OIE since 1994, Alex Theirmann is also a long-time employee of USDA/APHIS.

Immediately following the discovery of the third case of BSE, the USDA announced that it would reduce BSE testing. That's correct. Reduce testing, not increase. Out of 35 million cattle slaughtered each year, the USDA will now only test approximately 40,000 (a little more than one-tenth of one percent.)

In January 2004, shortly after the first case of BSE in the United States, the USDA accepted applications for any company that wished to establish their own USDA-approved laboratory for independent BSE testing. Creekstone Farms submitted an application and spent $500,000 building a state-of-the-art testing facility. While the USDA was reducing the number of animals tested for BSE, Creekstone Farms wanted to satisfy their customers' demands (particularly the Japanese market) and test every single animal they slaughtered, approximately 350,000 animals per year. But the USDA refused to allow Creekstone to purchase any testing kits and threatened the owners with imprisonment if they attempted. The USDA claimed that allowing Creekstone to do independent testing would "undermine confidence in the USDA's official position that random testing was scientifically adequate to assure safety." The USDA's justification? The obscure 1913 Virus-Serum-Toxin Act, a law intended to assure the safe supply of animal vaccines. Gregory L. Berlowitz, an editor at the *University of Illinois Law Review*, called the USDA's action against Creekstone "a ruse to protect the agency and the beef industry from a public outcry that would take place if more cases of mad cow disease were found. In its ninety-three years of existence, the Virus-Serum-Toxin Act has never been used or interpreted to regulate testing of any kind. Manipulating the act to include the BSE test perverts the statute's purpose." Berlowitz added, "The net result is that the USDA has placed the welfare and promotional concerns of the beef industry ahead of public welfare."

On March 23, 2006, the five-year anniversary of the seizure of our sheep and the end of the federal quarantine on our farm, Creekstone Farms launched a lawsuit against the USDA challenging the USDA's claim of exclusive rights to the BSE testing kits.

The USDA not only put corporate agriculture concerns ahead of public welfare in the United States, but in January 2006 they also applied pressure on Japan to reopen the import market, despite the reduction in BSE testing and the beef shipments sent to Japan that contained illegal specified risk materials.

Japan capitulated to U.S. demands and announced their intention to reopen the border to US beef imports in May 2006. Japanese consumers were outraged, and half of the Japanese mad cow advisory panel resigned in anger. One former panel member, Morikazu Shinagawa, told the *Dow Jones Newswire* that "he 'couldn't continue to work' on the panel because the conclusion to resume imports was preordained by the government." The upcoming Japan–U.S. summit fueled speculations that the decision to reopen the trade market was politically based, and *The Japan Times* wondered if the resumption of importing American beef was simply "a political gift to aid Mr. Bush. Both the U.S. administration and Congress are feeling the industry's pressure, something that needs to be relieved before the U.S. mid-term elections in November."

Whatever happened to a government that was "of the people, by the people, for the people"? How did big business and lobbyists take control of our government? What can be done about it? After decades of political wrangling, misuse and abuse of science, and catering to the demands of lobbyists and corporations, the USDA is now a corrupt behemoth that needs to be dismantled and restructured. Control of agriculture should be in the hands of local government.

The federal government may be required for regulating food in cases of interstate commerce and shipments across national borders, but otherwise the feds should stay out of the food business. The USDA has lost its credibility in ensuring a safe, nutritious food supply. The power of the federal government (and the corporations who pull its strings) has usurped the protections that state government should provide us.

The best—and perhaps the only—way to make a change is by voting with our dollars. Elections can be manipulated, but spending our money on products from people with ethics we share will ultimately have the biggest impact. Keep your dollars in your local community. Farmers markets are spreading throughout the country and are a great way to find farmers, talk with them, and buy their products. Many areas now have CSA (community supported agriculture) farms—another great way to get involved. When people develop personal relationships with their local farmers, it results in stronger communities, a healthier local population, a healthier ecosystem, and decreased dependence on fossil fuels.

Despite the ongoing eight-year battle with the USDA, our family's commitment to organic, sustainable agriculture is stronger than ever. We still sell products from local farmers at Schoolhouse Market. Rootswork now has a community radio station (WMRW), and Walt Krukowski continues to grow sustainable cut flowers. Larry and Jackie (now a sophomore at Middlebury College) are still making raw milk cheeses and teaching cheesemaking courses, and Heather (now a junior at Middlebury College) takes time each summer to manage the store and a highly successful booth selling Vermont cheeses at the local farmers market. Francis graduated from St. Lawrence University, and he helps out on the farm and at the store while cultivating his own construction and painting business. The community gardens behind the schoolhouse are thriving, and we are blessed to have Sally, Anda, and Jeremy Gulley growing wonderful organic vegetables on the farm.

Robin McDermott and I started a "Localvore" movement here in the Mad River Valley to educate people about our local farmers and their products, and to help our community strengthen its local food system. As Joel Salatin said, "A community which can feed itself is free."

And where is Linda Detwiler? She is now a consultant to Wendy's and McDonalds.

LINDA FAILLACE
July 2006

ACKNOWLEDGMENTS

It takes a village to raise a child," and it took an entire community to bring this book to fruition. One of the first things people mention after hearing our story is the fact that Larry and I are still smiling, still happy, still in love. My response is always the same: we could not have survived without each other, our family, our friends, and our incredible community.

Professor Eric Lamming was responsible for my study of TSEs, and I am thankful for the chance to work and learn from him. He passed away in June 2005 and is missed very much.

When we had the far-fetched idea to import sheep from Europe, Mr. and Mrs. Houghton Freeman shared our dream of building a sheep dairy industry and believed in us. The Freemans' amazing work aids Vermonters in a multitude of ways, often with little recognition or fanfare. Their work will benefit the Green Mountain State long into the future. Thank you for your vision, incredible generosity, and compassion. We are forever grateful!

Dr. Bernard Carton was responsible for helping us find the best possible sheep and for keeping us updated on the latest European research. I appreciate his wonderful sense of humor and his friendship. Thank you, Bernard!

A very special thank-you to all the Dutch and Belgian shepherds—the DeVleigher family, Mr. and Mrs. Blanchaert, Mr. and Mrs. Pississersens, Mrs. Kostense, Mr. and Mrs. Tylleman, Mr. and Mrs. Atema, Ms. Porte, Ms. Brouwers, and many more—who welcomed us so warmly into their homes and provided us with our beloved sheep. And our trips to Ghent would not have been complete without visits to our favorite artisanal cheesemonger, Michel Peeters, and his beautiful wife, Françoise.

Several people within the USDA were extremely helpful, including Dr. Roger Perkins, Mark Dulin, Dr. Jim Crawford, Dr. Mary Jo

Schmerr, Dr. Clarence "Joe" Gibbs, Dr. Katharine O'Rourke, and Dr. Jon Hansen.

The Belgian Embassy and the Belgian Ministry of Agriculture were also very helpful, and special thanks go to Mr. Jaak Gabriels, Ms. Danielle Borremans, Dr. Maillet, and Mr. Adriansens.

We appreciate all the scientific assistance we received from Dr. Piet Vellema, Dr. Shu Chen, Dr. Glenn Telling, Dr. Bruno Oesch, and Professor Vanopdenbosch. Mark Purdy is a visionary who has gone against the grain to pursue alternative theories (on a shoestring budget). Mark's efforts bring the scientific community one step closer to finding the cause and possible cures for BSE. And a special thank-you to Dr. Tom Pringle for all his advice and encouragement. You helped me believe in myself. (The movie was your idea first!)

Raising sheep in Vermont was made easier by the Connecticut River Valley Partnership program, Dr. David Henderson, Dr. Roger Ives, Dr. Karen Anderson, and the support of our fellow shepherds, including Jim McRae, Chris and Barbara Hall, Laini Fondiller, Bambi Freeman, Betty Berlenbach, Dan and Daphne Hewitt, Elizabeth and Ken Squier, Tod Murphy and Pam VanDeursen, Linda Doane, Stewart Skrill, Chet Parsons, Willow Smart, Doug and Sarah Flack, and Harvey Warrick. Our importation from New Zealand would not have been possible without the professional staff at Silverstream, including Dr. Jock and Hilary Allison, their daughter, Katharine, and her husband, Andrew. Thank you for your warm hospitality.

Freddie Michiels and his very talented wife, Lutgard, are responsible for our cheesemaking success, and we consider them family. Freddie was there for us every step of the way and was incredibly supportive, particularly in July 1998 and July 2000. You kept us going "even when the bombs were dropping"! Thank you, Freddie.

Barry Simpson is an amazing statesman. His wife, Claire, provided us with savory sustenance at critical times in the battle against the USDA, and their daughter Sarah will be an incredible contributor to future TSE research. Barry and Claire, your persistence, unwavering pursuit of truth, and fearlessness have kept us going, and we are extremely grateful.

The Mad River Valley is blessed to have George and George Schenk. If anyone is ever in need, the Schenks and American Flatbread are the first to respond. With their commitment to quality local food, support of the community, and dedication to small family farms and earth stewardship, I am proud to be considered their friend.

A very special thank-you to Edie Connellee and Bill Carnright for never backing down, and for picking us up and encouraging us to keep going at times when we felt overwhelmed. You taught us to be strong.

Thank you, Anne Burling, for your commitment to organic agriculture, creation of Rootswork, and incredible patience and benevolence.

Thank you, Bruce, for being my partner at the store and for your tremendous compassion. Jacki, Marc, and Laci Harmon were instrumental in helping us reopen the store in 2001 and taking care of things when the going got crazy. Duke and Laurie McGuire made sure Pélé was always in good hands and watched over the family too. I look forward to finally taking our trip together, Laurie.

When the USDA came after us, the following people created a safe web around us and kept us fed, nurtured, and sane: Brad (who fell in love with Kanga), Terry Allen, John and Devo Barkhausen, Virginia Barry, Heidi Benjamin, Rupert Blair, Kortney Bryant, Millie Chapell, Jito and Bonnie Coleman, Ross Conrad, Keith Davidson, Dee from Massachusetts, Richard Denby, Jim Edgcomb, the Eno family, Steve Farnham, Mattye Hard-Douglas, Graeme Freeman, Don Green, Cory Hatch, Jim Hoag, Ted Joslin, Ruth Joslin, Charles Kettles, Harriet and Dick King, Phil and Onrie King, Joe Kline, Walt Krukowski, Alan LaPage, Bernie Lewis, Jim Leyton, Suzanne Lupien, Pam McBrayne, Lisa Mccrory, Joannie, Will, and Rick McGraw, Anne Miller, Roy Morrison, Wesley and Charles Porter, Radkin, Joan Rae, Hanna Schenk, Brett Seymour, Sue and John Simms, Dorothy Singleton, Nancy and Ian Spencer, Ellen Strauss, Dorothy Tod, Brian Tokar, Martin and Kelly Von Trapp, Anna Whiteside, Callie Willis, Mary Williams, Steve Lidle and Nancy Wright, and the staff at American Flatbread, Lareau Farm, the Den (especially Nicole), and Cooking from the Heart. Thank you to Anson, Terry, and everyone at the Warren post office for delivering all the mail, even those only addressed to "the sheep people in Vermont." And a special thank-you to Ed Dooley for his amazing filmmaking skills and to Rick and the Mad Mountain Scramblers (John Bridgewater, Jayson Fulton, Ben, and Hollace Pratt) for their wonderful soundtrack for the documentary. And thanks to the many others who contributed in ways unbeknownst to us but appreciated.

Rural Vermont is sometimes the lone voice for small family farmers and sustainable farming in Vermont politics. We were lucky enough to have personal support from Alexis Lathem, Ellen Taggart, Ron Morrissette, Amy Shollenberger, and Dexter Randall.

When Francis received his internship at Harvard and needed a place to stay, Dr. Jim Hickey was a total stranger, and yet he generously gave Francis the use of his home, car, bike, cell phone, and dinners at some of Boston's finest restaurants. We are extremely thankful and anxious for Jim to move to the valley full time and open his brew pub!

Vermont has always had a reputation for free thinking and local government that supports people's rights. The following elected officials did not hesitate to stand up against the power of the federal government in our defense: Bruce Hyde, Jeb Spaulding, Kinny Connell, Ann Cummings, and the Town of Warren Selectboard. Thank you!

One of the benefits of our struggle was the opportunity to meet Darryl and Anthea Archer. They fought the same battle with tremendous poise and welcomed us into their home and community. May your family and farm long prosper!

The world would be a better place if there were more lawyers like Davis Buckley. Davis was incredibly supportive and generous with his time. We hope to get to spend more time with Davis and his family—but out of the courtroom!

When I first wrote the book I had five pages written exclusively about Roger Hussey and his influence on our family's lives. One of the most unselfish people I know, Roger quit his job to establish Save Our Sheep and used his entire savings to fund the documentary. He went out of his way to give individual attention to the children and keep the family whole. Roger is one of my guardian angels.

Writers across the country benefit from the work of Charlotte Dennett and Gerard Colby, head of the National Writers Union. They are responsible for encouraging me to write this book, and they provided some literary assistance. Once I began writing the book, Shannon Gilligan held my hand (and edited my work) for the entire process, while Ray Montgomery cheered me on. In Shannon I found a friend and a teacher. Whenever I hit what felt like an insurmountable problem, she would help me find a way around it. Shannon believed in me and made me feel anything was possible—it was. With love and lots of hugs, I thank you!

When I married Larry I was welcomed with open arms into his loving family, and they have helped us in many ways. Mom and Dad, you literally gave us shelter from the storm. Mark, Aileen, Nick, Jacob, and Lucas, it is reassuring to know that a supportive family is pulling for us—we love you!

Thank you to Bruno for your continual encouragement, especially every time you found me working late. And a special thank you to Julie, Paul, and Ellen and the rest of the Three Mountain Café crew who kept me well fed and caffeinated throughout the writing process. (Thanks for continuing to talk to me during baseball season, Ellen!)

Every writer dreams of finding the perfect publisher, one who truly believes in her book. I consider myself one of the lucky few. From the moment I first met Margo Baldwin and Ben Watson from Chelsea Green I knew I was in good hands. Helen Whybrow is an incredibly talented editor, and this book is stronger because of her input. Thanks also go to Marcy Brant, Jonathan Teller-Elsberg, Beau Friedlander, and the entire Chelsea Green staff for making this book the success that it is. And most of all thank you to John Barstow and Margo Baldwin for their integrity and determined pursuit of justice. Your positive actions help bring the changes needed in the world.

My love of writing came from one of my best friends, my father. You're my inspiration. And my strong will came from another friend, my mom. You're my keeper of dreams. I could not have asked for better parents and am still in awe of having such a wonderful family with four amazing sisters—Mary, Monica, Patty, and Becky. I love you all!

I have been blessed with many wonderful friends throughout my life but two people stand out like shining stars—Robin McDermott and Ray Mikulak. I have never experienced more generosity, friendship, and love than I have with the two of you, and I am extremely thankful for all the wonderful time we get to spend together. I am humbled to be your friend, and I love you both!

None of this would have been possible without my three incredible children and loving husband. You never cease to amaze me! I want to thank all of you for your encouragement, belief in the book, and, most of all, holding to your morals and never giving up. I am thrilled with the fact that I get to share my entire life with you all.

Francis is smart, perceptive, and a born leader. He has handled all of life's challenges with a sense of grace that goes way beyond his years.

Francis knows how to make things happen and pursues his dreams with incredible success. Thank you, Francis, for setting an example, for all your hard work and your incredible sense of humor, and most of all, for being my son.

Heather has a heart of gold and an insatiable curiosity, and she is the most tenacious person I know. She inspires me with the way she sets goals and keeps going regardless of any obstacles. Thank you, Heather, for your hugs, encouragement, editing, and for being my friend.

Jackie is a natural caretaker. Even as a young child, she was like a wise, quiet grandmother: always watching out for everyone, gently correcting and teaching. As a parent I try very hard to encourage all my children to make the most of themselves, find out what makes them truly happy and to go for it. I want the children to recognize the power of positive thinking and to see how much further ahead you can get in life with a great attitude. And I believe they have all mastered this, especially Jackie. Thank you, Jackie, for your incredible wisdom and abundant love.

Most importantly, I owe this book to the man of my dreams, Larry. My husband loves me with all his heart, returns all the love I give him and more. He's my best friend and treats me like a queen, spoils me with wonderful food, trips, and laughter, and is always there for me, ready to encourage me, hug me, or take me (sometimes on very short notice) to Montreal. Together we make our dreams come true. I love you with all my heart (and my cells).

INDEX

Veterinary Record, BSE article, 54, 74
Virus-Serum-Toxin Act of 1913, 304
Volunteer Scrapie Certification Program, U.S.
    See scrapie monitoring/prevention programs
VSBA. See Vermont Sheep Breeders
    Association (VSBA)

W

Wall Street Journal, 256–57
Walton, Tom, 85
Warren, Town of, selectboard's letter of support,
    111–12
Warren Meadows Cheese, 98, 99
water buffalo, possible TSEs in, 211–17,
    273–74
Waters, Mary, 256
Wells, Gerald, 81–83
Western blot tests, Rubenstein's, 123–26
Weybridge, England. See Central Veterinary
    Services Laboratory (Weybridge, England);
    Veterinary Laboratories Agency (Weybridge,
    England)
Whiteside, Anna, 23, 36, 67, 100

Whitten, Frankie and Marybeth, 27–28, 71,
    84, 191
    cheesemaking by, 96–97
    at Detwiler meeting, 62
Williams, Elizabeth, 193
Wisconsin Sheep Breeder's Conference, 17
World Health Organization, 108
World Organization of Animal Health (OIE),
    303–304

Y

Yestermorrow, 142–43

Z

Zeilenga, Wayne, 52, 57–58, 61–62, 280
    at Detwiler meetings, 53–56, 62–64
    disinfection of Faillace farm, role in, 265–68
    fear promoted by, 84
    formal quarantine of sheep, role in, 71–72
    importing sheep, role in, 12, 33
    scrapie inspection by, 53, 200
    seizure of sheep, role in, 231
    sheep appraisals, role in, 149
    warning about USDA, 72